# World Disasters Report

## Focus on reducing risk

International Federation
of Red Cross and Red Crescent Societies

2002

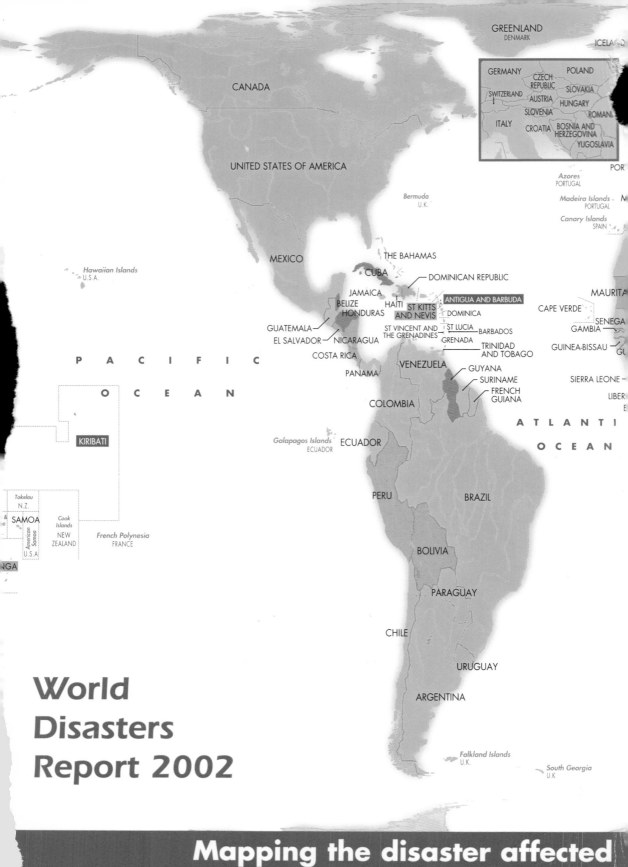

# World Disasters Report 2002

## Mapping the disaster affected

percentage of populations killed and affected by natural and technological disaster

# Contents

## Section One Focus on reducing risk

# Section Two Tracking the system

# Risk reduction is everyone's business

This year marks the tenth anniversary of the *World Disasters Report*. On the positive side, the past decade has seen a drop in the numbers of people killed by disasters. In the 1970s, natural disasters alone claimed nearly 2 million lives – by the 1990s this had fallen to under 800,000. But this is still a terrible and premature loss of life. Meanwhile, those affected – whether left injured, homeless or hungry – tripled to 2 billion during the past decade. Direct economic losses multiplied five times over the same period, to US$ 629 billion in the 1990s.

Our tenth report looks at how to reduce the risks that natural disasters pose to vulnerable communities around the world. The first report, in 1993, argued that the effectiveness of disaster response and the sound use of donors' money are "primarily dependent upon good disaster preparedness. All disasters are first tackled at the local level by local organizations. International response is built upon those local efforts."

This remains just as true a decade later. Disaster preparedness pays. When the most powerful hurricane for half a century hit Cuba in November last year, effective disaster planning and preparedness ensured that 700,000 people were evacuated to safety. When two years of record floods inundated Mozambique, well-prepared local and national resources saved 34,000 people from drowning. In 1999, of the 50,000 people trapped by Turkey's devastating earthquakes, 98 per cent were saved by local rescuers. So investing in disaster preparedness within communities at risk remains a top priority.

However, preparing to respond to disasters is only part of the broader risk reduction agenda. Where possible, measures to reduce the physical and human impacts of disasters must be taken. These mitigation measures take many forms. In Viet Nam, for example, a combination of embankments and specially-planted coastal mangrove forests protect vulnerable shores from storm surges and flooding. In Europe, where earthquakes have killed more people in the past decade than all other disasters put together, ensuring that construction codes are enforced is essential.

Physical protection must be complemented by better information at all levels. Often, exposed communities have expertise in dealing with risks which could be shared more broadly. In India, for example, traditional rainwater harvesting has helped thousands of households in combating drought. Equally, governments and aid organizations play a key role in promoting greater public awareness of disaster risks and how to deal with them. Enormous willpower, resources and imagination are needed across all sectors of human society to reduce the threat of disasters before they strike.

Since the attacks that rocked the world on 11 September last year, some global leaders have argued that fighting poverty will help promote a safer world. In March this year, donor nations committed more resources to achieve the international development goals of 2015, which include halving poverty and hunger, combating infectious diseases, and ensuring universal primary education.

Disasters, however, can wipe out years of development in a matter of hours. Big one-off disasters destroy farmland, animals, livelihoods – keeping people poor and hungry. Small recurrent disasters wear down family resources and resilience, exposing people to disease and poor health. Children may lose the chance to be educated if a disaster demolishes their school, or if parents need their help rebuilding shattered family lives.

So reducing the risks posed by disasters is not an optional extra – it is central to the very success of development itself. Disasters threaten to derail progress towards 2015's development goals. And if development remains blind to these risks, the chances of disaster will increase. Poverty is not the only reason why communities are exposed to disaster. As Cuba's experience shows, you don't need to be rich to be well informed and well prepared.

With this in mind, I would like to leave you with one thought. The international development targets are of enormous use in concentrating the minds and resources of governments and communities alike. To these we must add disaster risk reduction targets. Such as halving the numbers killed and affected by disasters, increasing the number of governments with dedicated disaster preparedness plans and resources, and boosting the amount of emergency and development aid spent on disaster mitigation and preparedness.

This anniversary edition of the *World Disasters Report* provides powerful evidence that investing in preparedness and mitigation helps combat the terrible human and economic toll of disasters. Reducing disaster risk is an urgent priority not only for disaster managers, but for development planners and policy-makers across the globe.

Didier J Cherpitel
Secretary General

chapter 1

Section One

**Focus on reducing risk**

# Risk reduction: challenges and opportunities

Hazards are an unavoidable part of life. Each day, every one of us faces some degree of personal risk from hazards of one kind or another. The hazards we face are very diverse. They arise from our society (for example, conflict, terrorism, civil strife) and technology (industrial and transport accidents), as well as from the natural environment (such as floods, windstorms, droughts, earthquakes) and from threats to public health.

Risk, and how we manage it, has become the subject of increasing research and debate in recent years. Is this heightened interest just a passing trend, or does it signify that the world has become a more dangerous place? Are the earth's 6 billion inhabitants becoming more vulnerable to disasters?

Risk reduction – why it is needed, how best to go about it and the challenges we face in achieving it – is the theme of this *World Disasters Report*. The report focuses on the opportunities and challenges we face in reducing the risks from natural hazards. This emphasis does not imply that the threats from conflict and political violence are insignificant. But such "complex political emergencies" have received much attention during the past decade, while the importance of what are often (erroneously) called "natural" disasters has been overlooked.

In this chapter, we outline the scale and nature of the threats posed by disasters and some of the obstacles to effective risk reduction. We demonstrate how disaster mitigation and preparedness can pay dividends in terms of lives and livelihoods saved. And we conclude that:

- disaster mitigation and preparedness must form part of the wider context of risk reduction – relevant to all those working in hazardous regions, whether in relief, development, business, civil society or government;
- long-term partnerships based on good governance across many sectors and disciplines provide the best basis for tackling the threats posed by disasters; and
- setting targets for risk reduction could provide a way to focus political will and adequate resources on the problem.

## Disasters threaten development...

Data gathered worldwide over the last three decades by the Brussels-based Centre for Research on the Epidemiology of Disasters (CRED) and reinsurance giant Munich Re suggest that, while the number of people killed by natural disasters has levelled out at around 80,000 per year, the number affected by disasters and associated economic

In Orissa, India, preparedness drills are organized at local cyclone shelters to ensure that communities know what to do when storms strike.

Orissa Disaster Mitigation Programme, India 2000.

losses have both soared. During the 1990s, an annual average of around 200 million people were affected by natural disasters – nearly three times higher than during the 1970s. Economic losses from such disasters in the 1990s averaged US$ 63 billion per year – nearly five times higher in real terms than the figure for the 1970s (see Figure 1.1). Even this figure is dwarfed by an estimate from Munich Re's leading geo-scientist, Gerhard Berz, that global warming-related disasters could soon cost over US$ 300 billion every year (see Chapter 4, Box 4.5).

While these figures sound sobering enough, they disguise the devastating effects that disasters can have on poorer nations' development. According to the Honduran prime minister, Hurricane Mitch, which killed up to 20,000 Central Americans in 1998, put his country's economic development back 20 years. In the same year, Peru suffered El Niño-related storm damage to public infrastructure alone of US$ 2.6 billion – equivalent to 5 per cent of gross domestic product (GDP). Losses from 1999's earthquakes cost Turkey around US$ 20 billion, and losses due to landslides in Venezuela in the same year cost US$ 10 billion – both figures are equivalent to over 10 per cent of each nation's GDP.

Most of these cost estimates are based only on damage to physical infrastructure. So these statistics tell only part of the story. The costs inflicted by disasters on human livelihoods

## Thirty years of "natural" disasters

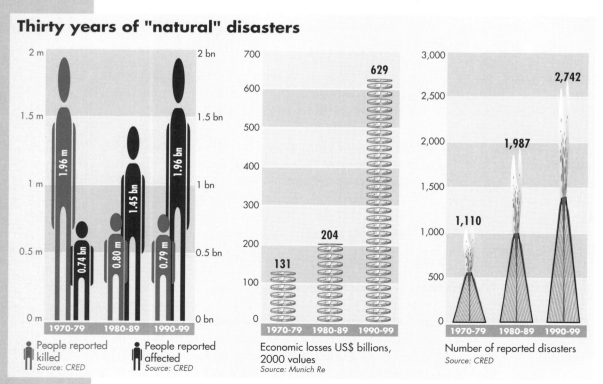

People reported killed
Source: CRED

People reported affected
Source: CRED

Economic losses US$ billions, 2000 values
Source: Munich Re

Number of reported disasters
Source: CRED

and social capital – costs paid by individual families and communities – are almost incalculable. As a Mozambican woman said after 2000's floods: "We lost everything we had worked for during our lives. We do not know when and where to start."

Furthermore, even the most reliable disaster statistics cannot reveal the full impact of natural hazards on society, because many hazard events fall below the arbitrarily defined threshold of what constitutes a "disaster".

To get a better picture, researchers in Latin America have been collecting data on the direct impact of all events involving natural and technological hazards in 13 countries in the region. The findings from the research, stored in the DesInventar database, are alarming: they show that the total impact of what the researchers call "everyday disasters" in some of these countries may be much greater than that of the much smaller number of large events that are formally recorded as disasters. For example, data from one country, Guatemala, covering the period 1988-1998 (excluding Hurricane Mitch) records 1,666 individual events leading to 1,393 deaths and 395,961 people affected. Over the same period (including Mitch), CRED's EM-DAT database (the source for disaster data in Chapter 8 of this report) records only 19 disaster events in Guatemala leading to 859 deaths and affecting 192,830 people.

It is well documented that the poor are usually those hardest hit by disaster. They are more likely to occupy dangerous, less desirable locations, such as flood plains, river banks, steep slopes and reclaimed land. Lack of financial and material resources gives poor people less flexibility in protecting their livelihoods and homes against disaster. When disaster strikes, assets bought with loans (for example, a sewing machine or a cow) can be instantly destroyed. This forces the poor back to the beginning – or worse, since they have to pay back the loan for an asset long lost.

Disasters undermine development by contributing to persistent poverty. As Didier Cherpitel, secretary general of the International Federation, says: "Disasters are first and foremost a major threat to development, and specifically to the development of the poorest and most marginalized people in the world. Disasters seek out the poor and ensure they stay poor."

## ...but flawed development drives disasters

However, vulnerability to disaster is determined not simply by lack of wealth, but by a complex range of physical, economic, political and social factors. Flawed development is exacerbating these factors and exposing more and more people to disasters.

While population growth and rapid, unplanned urbanization force poorer groups to live in more hazardous areas, even the better-off are at risk: inhabitants of the

Figure 1.1 (opposite page): "Natural" disasters include floods, windstorms, droughts, earthquakes and volcanoes, but not epidemics or technological accidents. The word "natural" is in quote marks to reflect the reality that, while the hazard which triggers the disaster may be a natural phenomenon, the root causes of the disaster may often be unnatural in origin.

apartment blocks that collapsed during earthquakes in Turkey in 1999 and Gujarat in 2001 were victims not of poverty, but of ineffective building codes (see Chapter 5). And it was middle-class housing that was swept away by the Santa Tecla landslide in San Salvador in January 2001.

In the Vietnamese city of Hue, the expansion of infrastructure over the past decade – including bridges, railway lines and roads – has created a barrier across the valley. Excess rainfall can no longer soak away quickly, and floods in the area have become more severe as a result (see *World Disasters Report 2001*). Likewise, the growth of infrastructure across the globe has increased both the level of assets at potential risk from disasters, and the numbers of people dependent on such lifelines as electricity, gas and water mains.

Economic growth does not necessarily imply a reduction in risk, particularly in the lowest-income countries. Economic pressure can bring environmental degradation; deforestation, in particular, has disrupted watersheds, leading to more severe droughts as well as floods. Social and economic changes can undermine traditional extended family structures, once an important form of support during crises. People switch jobs or change the type of crops grown in response to improved marketing opportunities, but in so doing, they may increase their vulnerability to disasters. The level of community solidarity and the extent of a community's participation in decision-making can also affect exposure to disaster risk.

Finally, the effectiveness of community and government arrangements for disaster preparedness and mitigation are also critically important. Such arrangements include the extent of flood-proof dykes; early warning systems against droughts, floods or cyclones; evacuation routes and shelters; stockpiles of relief materials; well-trained and coordinated disaster response teams; and disaster-aware populations.

Clearly, disasters are a major threat to the global economy and to society. The old view of disasters as temporary interruptions on the path of social and economic progress, to be dealt with through humanitarian relief, is no longer credible. Nor can a simple link be made between reducing poverty and reducing disaster. The problem is much deeper; it stems from fundamental flaws in the development process itself. Sustainable development is society's investment in the future. That investment will be squandered if it is not protected adequately against the risk of disaster.

## Barriers to more effective risk reduction

While there is little doubt that disasters pose grave threats to lives, livelihoods and development in many hazard-prone nations, there remains a lack of commitment to risk reduction. Some reasons why include:

**Geopolitics.** The 1990s saw a series of major conflicts across the world, for example in the Gulf, Somalia, former Yugoslavia, Rwanda, Sudan, Chechnya and East Timor. The explosion of these complex political emergencies, and the massive human suffering that resulted, dominated the humanitarian agenda – using up vast amounts of international aid funds. The problem of vulnerability to natural hazards was set aside. Recently, Hurricane Mitch and the Bangladesh floods in 1998, the Orissa and Mozambique cyclones in 1999 and 2000, and earthquakes in Turkey in 1999 and Gujarat in 2001 have drawn aid agencies' attention back to natural disasters. But events since 11 September 2001 illustrate how quickly international priorities can shift in response to political crises.

**No coherent risk reduction "community".** Those professionals trying to prevent disasters and deal with their consequences come from a range of disciplines (natural and social scientists, engineers, architects, doctors, psychologists, development and humanitarian workers) and organizations (international aid agencies, governments, civil society organizations, academics, consultants and businesses). All too often the "disaster community" is characterized by fragmentation along disciplinary and institutional boundaries.

**Risk reduction is seen as a separate sector.** In reality, however, it is a cross-sectoral issue. It falls into the gap between humanitarian assistance and long-term development cooperation, when it should be part of everyone's work. It is implemented as an occasional add-on to other projects, when it ought to be mainstreamed into development and humanitarian programming and approached as a long-term process. As a result, hazard risks are rarely considered in the formulation of development projects or broader policy. Instead, risk reduction concerns are either marginalized, at best, or simply forgotten.

**Risk reduction is still viewed as a technical problem.** Since the 1970s, the relationship between human actions and disasters has been increasingly well documented and argued. By the mid-1990s, the significance of social, political and economic vulnerability to disasters was widely accepted within academic circles. But this new thinking has so far proved ineffective in breaking down barriers between disasters and development at *operational* level. Most mitigation efforts still address the visible signs of vulnerability, such as poor housing and unsafe locations. Because these are seen as physical or hazard-related problems, they are addressed mainly through technical solutions, such as embankments against floods or improved construction against earthquakes. Meanwhile, the underlying factors that compel people to live in insecure conditions remain unaddressed.

**Lack of resources committed to risk reduction.** Although agencies' policy statements may suggest there is widespread concern about risk reduction, the acid test of the international community's commitment is surely the amount of aid funding committed to it. Here, the picture shows that there is a lot of room for improvement. While donors are usually quick and generous in post-disaster relief and reconstruction, they dedicate far fewer resources to risk reduction. A few, notably the European Union (EU) and the United States' Agency for International Development (USAID), have established dedicated mitigation and preparedness budget lines under their broader disaster management programmes. But expenditure is very limited. For example, the European Community's Humanitarian Office (ECHO) set up a separate disaster preparedness budget in 1994 (which later became the DIPECHO programme). However, while DIPECHO spent 8 million euros (US$ 7 million) last year, this represents just 1.5 per cent of the total ECHO budget for humanitarian aid.

**Invisibility of risk reduction spending.** Some expenditure also occurs under donor development programmes. This may involve projects clearly identifiable as mitigation or preparedness. Or it may include mitigation as a design feature of another project (e.g., hurricane-proofing school buildings as part of an education project). However, this expenditure, whether or not it involves a dedicated project, is rarely, if ever, reported in donor accounts. So it is impossible to determine total expenditure on disaster mitigation and preparedness – a revealing fact, which implies that such spending is of little political interest. Nor can we assume that this expenditure is particularly high, especially as most of those working in development view disaster mitigation as the responsibility of their counterparts in emergency units.

## Proving that mitigation and preparedness pay

A further obstacle is that we still have much to learn about how best to reduce vulnerability – what works, what doesn't and why? The subject needs more systematic study. Too often, the literature is dominated by superficial coverage and agency propaganda.

Monitoring and evaluation of risk reduction tends to be short term, and tied to donors' project cycles. It focuses on the outputs of initiatives (e.g., numbers trained in disaster planning, area sown with drought-resistant seeds), rather than their *impact* (e.g., extent to which lives and assets are better protected during disasters, improvements in food security during drought).

This is not to say that disaster mitigation and preparedness don't work. Far from it. Those who work in the business and have seen the results at first hand, believe very firmly that it does. There are many documented success stories from around the world – initiatives addressing different aspects of risk reduction, against a range of hazards and in a variety of ways. For example:

- The Bangladesh cyclone preparedness programme has successfully warned, evacuated and sheltered millions of people from cyclones since its inception in the early 1970s (see Box 1.1).
- In four drought-affected Indian states, the building and restoration of rainwater-harvesting structures (check dams, and community and household tanks) has rejuvenated local water courses and helped an estimated 20,000 villages to grow crops and maintain domestic water supplies.
- A study of 1,800 farm plots in three Central American countries hit by Hurricane Mitch demonstrated that farms using "agro-ecological" methods to prevent soil and water run-off from hillsides, lost far less topsoil, retained more moisture and were much less vulnerable to surface erosion than plots farmed using more conventional methods.
- When floods struck Viet Nam in 1999, only one out of 2,450 flood- and typhoon-resistant homes, built by the Red Cross, succumbed.
- Only weeks after setting up emergency response teams, a community organization in a Filipino village rescued 31 families from rising flood waters.

Overall, though, the picture is more patchy. Far too many initiatives are not properly evaluated or documented. As a result, policy-makers and planners often lack adequate information to decide about the most suitable approaches to risk reduction in different contexts.

There is an additional, inherent, problem in trying to prove that mitigation and preparedness pay. United Nations (UN) secretary-general Kofi Annan puts it succinctly: "While the costs of prevention have to be paid in the present, its benefits lie in a distant future. Moreover, the benefits are not tangible; they are the disasters that did *not* happen."

Ideally, natural hazard risks should be assessed at the appraisal stage of all potential projects in hazard-prone areas. Although environmental impact analysis will often explore the technological hazard risks posed *by* a project, an analysis of the risks *to* the project posed by natural hazards is seldom made. There is considerable reluctance among many governments and international organizations to do so, other than for structural mitigation projects such as dykes or sea defences.

The failure to take hazard risks seriously partly reflects budgetary constraints, particularly in the face of shorter-term and more pressing demands on funds. For instance, a joint USAID-Organization of American States (OAS) initiative, the Caribbean Disaster Mitigation Program (CDMP), found that "the most common response from the political directorate in the Caribbean to programs that would reduce vulnerability through more stringent building and development standards remains: 'Our nation is too poor to afford the required standards'. The challenge of

# Box 1.1 Cyclone preparedness saves millions in Bangladesh

During the 1990s, Bangladesh was lashed by five enormous cyclones. Up to 140,000 people died, most of them during one storm in 1991. But over 2.5 million people were evacuated – and their lives almost certainly saved – before the cyclones struck (see figure below). This was largely thanks to the cyclone preparedness programme (CPP) initiated in the early 1970s by the International Federation, the Bangladesh Red Crescent Society (BDRCS) and the government of Bangladesh.

The CPP was started after almost 500,000 people died during a cyclone in November 1970. Wind speeds reached 220 kilometres an hour and the tidal surge topped ten metres that year. In 1991, wind speeds were even higher, the maximum reported tidal surge was six metres and about 140,000 people died – but 350,000 were safely evacuated. In May 1997, a similar cyclone with winds of over 230 kilometres an hour and a tidal surge of up to 4.5 metres claimed less than 200 lives, while a million people were evacuated into shelters. Over the same period, the CPP was progressively extending its shelters and communications systems. Clearly the investment paid off.

The CPP can now alert around 8 million people across the whole coastal region, of whom it can assist around 4 million to evacuate. The warning system uses Asia's largest radio network, linking the CPP's Dhaka headquarters with 143 radio stations. Radio warnings are then relayed by 33,000 village-based volunteers using megaphones and hand-operated sirens. The volunteers are also trained to rescue people and evacuate them to shelters, administer first aid and assist in post-cyclone damage assessment and relief.

Between disasters, volunteers organize simulation exercises and disaster-awareness rallies. This vast network of volunteers helps the CPP to solve two problems which challenge other disaster mitigation and preparedness initiatives: the transformation from a "project-based" approach into a long-term, ongoing operation; and scaling-up from a community-based initiative to a system large enough to protect all those living in areas at risk.

The Bangladeshi government contributes 56 per cent of CPP's operational costs, which amounted to US$ 460,000 in 2001, and the International Federation covers the remainder. Local communities do not contribute to CPP's running costs, but raise funds to manage and maintain cyclone shelters. In total, there are 1,600 shelters across the coastal region, 149 of them built by the BDRCS. Each BDRCS shelter can accommodate up to 1,500 people and costs around US$ 78,000 to build, plus an annual maintenance fee of about US$ 780. Assuming the shelters last approximately ten years, that amounts to a cost of less than US$ 6 per head for each year of protection. ■

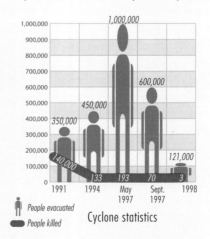

Cyclone statistics

People evacuated
People killed

the CDMP and similar programs that promote safer development consists in debunking this myth, by demonstrating that it is cost-effective to invest in mitigation of natural hazards."

The CDMP has demonstrated the potential benefits of structural mitigation through a retrospective analysis of public and private projects in the Caribbean which have suffered damage from tropical storms. The study concludes that the incorporation of hazard and vulnerability information into the earliest stages of project design or reconstruction is essential to ensure both hazard resilience and the lowest costs over the life of the project.

One of the projects considered by the study is a deep-water port in Dominica, constructed by the government to handle banana exports more efficiently and to lower the handling costs of imports. Work on the port facility lasted from 1974-78. A year later, Hurricane David hit the port, requiring reconstruction estimated at US$ 2.3 million (1975 prices) – equivalent to about 41 per cent of the cost of the original port. Had the original facility been built to a higher standard, able to resist David-force winds, it would have added only 12 per cent to construction costs. In such cases, clearly mitigation would pay, even defined in the narrowest, direct cost terms. The benefits of mitigation would be even greater if analysts factored in the indirect and secondary impacts of disaster – such as the economic cost of disruption to banana exports.

A key step in promoting broader hazard risk assessment is to compile a series of case studies demonstrating the benefits of mitigation and preparedness, building on the work of the CDMP. Techniques for analysing the special issues raised by disasters – such as potential loss of life and the probability of a hazard event of varying intensity occurring over the lifetime of a project – also need to be made more widely available. Project guidelines should be expanded to include assessing natural hazard risks and mitigation benefits.

The Dominica example above focuses strictly on physical investment in the context of an individual project. Yet a shift away from purely technical solutions and individual projects is essential. Successful risk reduction not only means disaster-proofing all vulnerable infrastructure, it requires an attitude of mind that permeates all echelons of society. For this to occur, greater appreciation of the long-term, socio-economic impacts of disasters, and better analysis of the benefits of mitigation and preparedness are required.

## International and regional commitment

Protecting against disasters is an enormous challenge. As we have seen, disasters are complex problems arising from the interaction between the environment and the development of human societies. They demand equally complex responses, drawing on a wide range of skills and capacities. Traditional disaster preparedness activities, such as improving early warning and evacuation systems, stockpiling relief supplies and strengthening disaster response capacities at all levels, clearly play a key role. However, because risk reduction needs to go to the heart of the development process, the challenge is well beyond the capacity of conventional disaster managers alone. It requires cooperation between multilateral development agencies, national and local government, non-governmental organizations (NGOs), businesses, natural and social scientists, technical specialists and, of course, vulnerable communities.

During the 1980s, this message seemed to be getting through. And with the establishment of the International Decade for Natural Disaster Reduction (IDNDR) in 1990, the member states of the UN committed themselves to reducing the impact of natural disasters through "concerted international action". Disaster mitigation and preparedness appeared to be firmly on the aid agenda.

Yet by the IDNDR's mid-term review in 1994 the international community was admitting to "the meagre results of an extraordinary opportunity given to the United Nations and its member states". Even at the end of the decade, Kofi Annan had to acknowledge that "the number and cost of natural disasters continues to rise" and there was still much to achieve. "We know what has to be done," he told delegates at the IDNDR's closing conference in Geneva in July 1999, "What is now required is the political commitment to do it." If the IDNDR did not live up to expectations, its successor, the International Strategy for Disaster Reduction (ISDR), has learned from that experience, and is based on more realistic expectations.

There are signs of growing interest elsewhere as well. Programmes relating to poverty alleviation are beginning to address vulnerability to natural hazards. The World Bank and UN Development Programme (UNDP) are actively addressing hazard risk. They have created special units to promote greater awareness among their own technical and geographical departments, and are supporting research, discussion and piloting of new approaches to vulnerability reduction.

The World Bank, supported by the International Federation and UNDP, has established the ProVention Consortium, a global coalition of governments, international organizations, academic institutions, the private sector and civil society organizations aimed at reducing disaster impacts in developing countries. The International Federation has issued guidelines on the "well-prepared National Society" to its 178 member societies. The OAS's work and the hemispheric summit process have initiated policies to encourage governments to deal with risk reduction as a development problem.

Figure 1.2 (opposite page): The lens of risk reduction. Risk reduction is everyone's business. Disasters pose too great a threat to development to remain the concern only of disaster managers and scientists. Risk reduction is a priority for policy-makers and field operators in every sector. From now on, planners have no choice but to view all their decisions through the lens of risk reduction.

Disaster preparedness (DP) and mitigation; coastal retreat; local coping strategies; adaptation funds; legal protection for migrants; international protocols; reduction of emissions

National DP plans and management; early warning; evacuation; stockpiles; agency coordination; public awareness; training; vulnerability and capacity assessment (VCA)

Coordination; quick, appropriate relief; local participation in assessment; strengthen local disaster response; relief as platform for recovery

**Disaster preparedness**

**Adaptation to climate change**

**Disaster response**

**RISK REDUCTION**

**Development**

**Disaster recovery**

**Disaster mitigation**

Mainstream risk assessment; strengthen livelihoods (human, social, political, financial and physical assets); sustainable agriculture and resource use; cross-sectoral partnerships; social services; diversified economies; good governance

Hazard-proof infrastructure, crops and jobs; building codes; retrofits; land-use regulations; insurance; micro-finance; public awareness; VCA; right to safety; targets

Assess risks during rehabilitation; local partners and procurement; livelihoods not just reconstruction; risk reduction advocacy opportunity

# Box 1.2 A centuries-old drought mitigation strategy

For the past two years, southern Sri Lanka has suffered a prolonged drought, described by locals as "the worst in 50 years". There has not been a successful crop in some areas for four or five consecutive seasons. Livestock has died, water in wells has dropped to dangerously low levels, children are increasingly malnourished and school attendance has fallen. An estimated 1.6 million people have been affected.

Muthukandiya is a village in Moneragala district, one of the drought-stricken areas in the "dry zone" of southern Sri Lanka, where half the country's population of 18 million lives. Rainfall in the area varies greatly from year to year, often bringing extreme dry spells in between monsoons. But this drought has been much worse than usual. Despite some rain in November, only half of Moneragala's 1,400 tube wells were in working order by March 2002. The drought has devastated supplies of rice and freshwater fish, the staple diet of inland villages. Many local industries have closed down and villagers are heading for the towns in search of work.

The villagers of Muthukandiya arrived in the 1970s as part of a government resettlement scheme. Each family was given six acres of land, with no irrigation system. Because crop production, which relies entirely on rainfall, is insufficient to support most families, the village economy relies on men and women working as day-labourers in nearby sugar-cane plantations. Three wells have been dug to provide domestic water, but these run dry for much of the year. Women and children may spend several hours each day walking up to five kilometres to fetch water for drinking, washing and cooking.

In 1998, communities in the district discussed water problems with a Colombo-based NGO, Intermediate Technology Development Group (ITDG) South Asia. What followed was a drought mitigation initiative, based on a low-cost "rainwater harvesting" technology already used in Sri Lanka and elsewhere in the region. It uses tanks to collect and store rain channelled by gutters and pipes as it runs off the roofs of houses.

Despite an indigenous tradition of rainwater harvesting and irrigation systems going back to the third century BC, policy-makers in modern times have often overlooked the value of such technologies, and it is only recently that officials have taken much interest in household-level structures. Government and other programmes have, however, been top-down in their conception and application, installing tanks free of charge without providing training in the skills needed to build and maintain them properly. ITDG South Asia's project deliberately took a different approach, aiming to build up a local skills base among builders and users of the tanks, and to create structures and systems so that communities can manage their own rainwater harvesting schemes.

The community of Muthukandiya was involved throughout. Two meetings were held where villagers analysed their water problems, developed a mitigation plan and selected the rainwater harvesting technology. Two local masons received several days' on-the-job training in building the 5,000 litre household storage tanks: surface tanks out of ferro-cement and underground tanks out of brick. Each system, including tank, pipes, gutters and filters, cost US$ 195 – equivalent to a month's income for an average village family. Just over half the cost was provided by the community, in the form of materials and unskilled labour. ITDG South Asia contributed the rest, including cement, transport and payment for the skilled labour. Households learned how to use and maintain the tanks, and the whole community was trained to keep domestic water supplies clean. A village rainwater harvesting society was set up to run the project.

To date, 37 families in and around Muthukandiya have storage tanks. Evaluations show clearly that households with rainwater storage tanks have considerably more water for domestic needs than households relying entirely on wells and ponds. During the driest months, households with tanks may have up to twice as much water available. Their water is much cleaner, too.

Nandawathie, a widow in the village, has taken full advantage of the opportunities that rainwater harvesting has brought her family. With a better water supply now close at hand, she began by growing a few vegetables. The income from selling these helped her to open a small shop on her doorstep. This increased her earnings still further, enabling her to apply for a loan to install solar power in her house. She is now thinking of building another tank in her garden so that she can grow more vegetables. Nandawathie also feels safer now that she no longer has to fetch water from the village well in the early morning or late evening. She says that her children no longer complain so much of diarrhoea. And her daughter Sandamalee has more time for school work.

In the short term, and on a small scale, the project has clearly been a success. The challenge for NGOs like ITDG South Asia lies in making such initiatives sustainable, and expanding their coverage.

At a purely technical level, rainwater harvesting is evidently sustainable. In Muthukandiya, the skills required to build and maintain storage tanks were taught fairly easily, and can be shared by the two trained masons, who are now finding work with other development agencies in the district.

The non-structural elements of the work, especially its financial and organizational sustainability, present a bigger challenge. A revolving fund was set up, with households that had already benefited agreeing to contribute a small monthly amount to pay for maintenance, repairs and new tanks. However, it appears that the revolving fund concept was not fully understood and it has proved difficult to get households to contribute. Madhavi Malalgoda Ariyabandu, head of ITDG South Asia's disaster mitigation programme, points out that "recovering costs from interventions that do not generate income directly will always be a difficult proposition", although this can be overcome if the process is explained more fully at the outset.

The Muthukandiya initiative was planned as a demonstration project, to show that community-based drought mitigation through rainwater harvesting was feasible. Several other organizations have begun their own projects using the same approach. The feasibility of introducing larger tanks is being investigated.

However, a lot of effort and patience are needed to generate the interest, develop the skills and organize the management structures needed to implement sustainable community-based projects. So it will probably be some time before rainwater harvesting technologies can spread rapidly and spontaneously across the district's villages, without external support. ∎

Meanwhile, researchers and disaster professionals have been developing their own synergies. The pioneering International Research Committee on Disasters, comprising a range of social science disciplines, was set up in the early 1980s. The International Geographical Union has recently established a Task Force on Vulnerability. A semi-formal Global Alliance for Disaster Reduction has been created by engineers, disaster managers and academics from all over the world, to document and promote good practice.

Influential regional networks for research, advocacy, education and training were formed during the 1990s, including Duryog Nivaran (South Asia); Peri-Peri (southern Africa); and La Red (Latin America). The Pan American Health Organization has a long history of involvement in this area, including strengthening hospitals to make them more disaster-resilient. And a few international development NGOs have identified risk and vulnerability as important issues in their policy statements.

Nevertheless, more commitment is needed. Even within those organizations that have begun to make changes, there are often strong internal battles to be won before risk reduction becomes a matter of concern for everyone in the organization and not just for the handful of professional staff involved.

## Lessons from front-line communities

Despite international initiatives, the front line against disasters is held by at-risk communities themselves, which are often the main actors in disaster mitigation and preparedness. Especially in developing countries, where the state's capacity to protect its citizens may be limited, communities rely on their own knowledge and coping mechanisms to mitigate against disasters, as they have done for generations.

When Hanna Schmuck (now the German Red Cross's disaster preparedness delegate to India and Bangladesh) studied the people who lived on the flood- and erosion-prone *chars* (silt islands) in the middle of Bangladesh's Jamuna River, she found their way of life was designed for maximum flexibility in an uncertain environment. Food, possessions and even houses could be moved out of harm's way at short notice. A local reed was used to stabilize new silt deposits and make them fit for cultivation. The agricultural calendar and crop varieties were planned to make the most of poor soil conditions and the annual flood cycle. Rice was sometimes planted in moveable seed beds before being transplanted when flood waters receded. Livestock kept by families for food and income could be moved away from flood waters. Marriage partners were sought on other *chars* and the mainland, to provide an escape route for relatives affected by floods. Local custom allowed families displaced by erosion to set up their homes on other people's land.

Community-based approaches to disaster mitigation are valuable for many reasons. They lead to a more accurate definition of problems and appropriate measures to overcome them, because they are based on vulnerable people's assessment of their real needs and priorities. They draw on local skills and expertise in living with disasters, which may be considerable. They can deploy low-cost, appropriate technologies effectively. They are more likely to be sustainable because they are "owned" by the community and build up local capacity.

Although there has been no systematic study of community-based mitigation and preparedness, there is a growing body of case-study evidence illustrating successful initiatives of many kinds. The Bangladesh cyclone preparedness programme depends for its effectiveness on the 33,000 community-based volunteers to relay radio warnings (see Box 1.1). An example of communities overcoming drought in Sri Lanka is illustrated in Box 1.2, while the challenges facing community-based disaster preparedness in Nepal are outlined in Box 1.3. Indeed, the value of community-based initiatives has led many disaster mitigation and preparedness programmes to abandon technical, top-down methods in favour of more participatory approaches.

For example, early warning systems for drought/famine and cyclones have benefited from the emphasis on social, as well as technical, priorities which community participation has brought. In the past, these systems tended to concentrate on improving scientific knowledge of hazards and forecasting of events. Nowadays, they put much greater emphasis on communicating warnings effectively and mobilizing communities through disaster awareness and training – as well as the more technical procedures of preparing safe areas for evacuation and stockpiling emergency materials. A growing number of drought/famine early warning systems are using community-level monitoring. Here, local people collect data on what they see as signs of growing food insecurity in their district, for example, an increase in the number of animals sold in local markets may indicate that families need cash to buy food because their own crops have failed. Such information is used in combination with satellite images and meteorological data to build up a fuller picture of the drought and its effects.

Participation has often meant simply including local people in implementing projects designed by outsiders. But participatory approaches should go much further, to give vulnerable groups greater control over the forces and institutions that influence their lives. One example is a community-based disaster preparedness (DP) programme in the cyclone-prone Cox's Bazaar district of Bangladesh. The programme has worked hard to ensure women are involved in the village DP committees responsible for maintaining cyclone shelters and transmitting warnings. But it has also tried to empower women more widely in their everyday lives through education and training in areas such as reproductive health, organizing self-help groups, and running small enterprises. Previously, women were not involved in DP work, and made up a high proportion of cyclone victims. Now, they are assuming a greater and more confident role. According to 40-year-old Shoba Ranishli, "Women definitely have to be involved in disaster preparedness, because women can then teach other women; men are not teaching women! ... Preparation for a cyclone at the household level is our work and responsibility. Men just tell what should be done ... but women just do it, we are more practical."

Paradoxically, disasters can be opportunities to bring communities together. In the village of Llhate in Mozambique, which was cut off by flood waters for two months in 2000, a group of older people formed their own association of elders to plough and plant the fields. "There is a palpable sense of solidarity," observed Albertino Cumaio, HelpAge International's emergency programme officer. "The simple need to survive, and the isolation, have made this community so united." Needing the help of younger people to work the fields, the elders took on household chores such as cooking and caring for children. This improved relationships between the generations: "We are gaining the trust and respect of the young ones through our contribution and the food we are producing for the community," said one of the elders, Mrs. Matusse. Trust, both within communities and between communities and government, is a key asset in the fight to reduce the toll of disasters.

The extent of local expertise and capacity can be exaggerated, however, and one must guard against romanticized views of the community. Skills and resources from outside are certainly valuable, provided that they are geared to local priorities and are deployed to enhance local people's capacities and not to marginalize them. Working closely with local people can help professionals to gain a greater insight into the communities they seek to serve, enabling them to work more effectively and produce better results.

## Good governance reduces disaster risk

The main weakness of community-based initiatives is their limited outreach. They may be very effective at local level, but how can they be scaled up to achieve greater impact? This is a major challenge. The community-based approach originated out of a realization that top-down, technology-driven models of mitigation and preparedness favoured by governments and international agencies were failing to make much impact in reducing vulnerability. The state, and its apparatus, were often seen as part of the problem. Yet scaling up is impossible without the participation of government at all levels and other influential actors.

It is generally agreed that national governments bear the prime responsibility for protecting their citizens against disasters. Some, such as Mozambique, take this more seriously than others (see Chapter 3), but it depends on their experience, resources and ideology.

The IDNDR played a significant role in encouraging governments to draw up national disaster management plans. However, the majority of existing plans focus on organizing emergency response, creating committees and listing governmental and civil responsibilities during disasters. National plans may mention longer-term mitigation and preparedness, but lack detail and dedicated resources. Disaster management is often viewed in a narrow, technical sense, rather than as part of a broader risk reduction strategy. Kofi Annan has warned that efforts to reduce vulnerability "often fail to engage the attention of top policy-makers" at national and international levels.

Mounting social and economic pressures, often coupled with policies favouring the reduction of state services, can undermine governments' capacity to reduce risks. For example, geographer Ben Wisner argues that El Salvador's disastrous landslides in January 2001 exposed the reluctance of a neo-liberal government to address key factors that it had earlier acknowledged as increasing vulnerability to disaster: inadequate public health services, insecure livelihoods, poor housing in unsafe locations, outdated government prevention and response structures, and a severely degraded environment.

Worse still, disasters can be exploited for political ends. Following earthquakes in Peru last year, an activity as seemingly neutral as risk mapping became a "political minefield" (see Chapter 2). The story of Peru's preparations for and responses to

Reducing the risk: women raise the embankment that protects their village from floods.

Peter Williams/World Council of Churches, Bangladesh, 2002.

## Box 1.3 Community-based disaster preparedness in Nepal

Nepal is no stranger to disasters. Half a million people were affected and 1,500 lost their lives when the catastrophic monsoon of 1993 triggered flash floods and landslides throughout the Himalayan foothills. The disaster hit the world's headlines. But smaller, recurrent disasters claim far more lives and often go unreported. From 1983-98, more than 18,000 people died during annual monsoons, landslides, fires, drought and disease epidemics. Many hill communities can be a week's walk from the nearest road head. If disaster strikes, emergency relief often arrives too late – or not at all.

Since the 1970s, the Nepal Red Cross Society (NRCS) has built up a network of relief warehouses and trained volunteers in disaster response. While aid stockpiles remain an important resource, the NRCS has, since 1997, increased its emphasis on community-based disaster preparedness (CBDP) in order to meet the challenge of recurrent disasters in remote communities. Dev Ratan Dhakwa, secretary general of the NRCS, explains: "There has been a shift from a traditional relief focus where communities are considered 'victims' and 'beneficiaries' of external aid, towards a more holistic and long-term approach that strengthens the communities' capacity to cope with hazards."

CBDP is centred around people, not technology. One of the Nepal programme's key objectives is to banish the sense of fatalism in the face of disasters which many communities feel and, instead, to motivate them to become more prepared and self-reliant, using the resources locally available. So far the programme has targeted 141 communities. A further 75 communities are planned for 2002-03, but the Maoist insurgency is hampering access to many hill villages.

A combination of leaflets, manuals and training sessions is designed to boost communities' confidence in their own capacity to cope. To date, the NRCS has trained 165 community volunteers who return to their villages and spread the disaster preparedness (DP) message. Training concentrates on first aid, basic rescue operations and tools such as hazard analysis, vulnerability and capacity mapping and contingency planning. Volunteers then work with their communities to map local risks and develop disaster plans. As well as preparedness measures, villagers may undertake small mitigation projects, such as building "gabions" (walls of stones wrapped in wire) or planting trees to prevent riverbanks and hillsides eroding.

Each monsoon, the river flowing through Daurali Panchkanya, a village in Parbat district, eats away more fertile farmland. Ganesh Rai, a voluntary DP trainer, recalls that flash floods once forced the entire community of 90 families to evacuate. But, he adds, "After the NRCS motivated us to undertake certain flood control measures, for which the village community contributed by its labour, a series of gabion boxes were constructed to check gully erosion. This has saved our crop fields." Planting trees along the riverbank and around homes has also helped reduce the damage from flood waters.

But monsoon floods are only the most high profile of a range of hazards facing Nepal's poorest communities. Open kitchen fires pose a constant threat. Singhya is a tiny hamlet on the plains of the *tarai*, targeted by the CBDP programme. According to Churamani Chowdhary, coordinator of the village development committee, "We launched an awareness campaign for fire control, on how to protect the thatched roofs from catching fire and being reduced to ashes in the process." The result has been spectacular, says Chowdhary. Against the previous annual average of 15-20 per cent of houses being damaged by fire, the village did not suffer a single case last year.

Women have played a prominent role in taking responsibility for risk reduction. In Parbat district, thanks to the initiative of village women, nearly 75 per cent of households in the four communities targeted for CBDP now have toilet facilities as part of the health and sanitation programme. This has drastically reduced the number of health-related problems. Meena KC, one of NRCS's female volunteers, says: "In each of the communities that I work with, women are in the forefront and are more active participants in various committees. This is because it is the women who have to bear the brunt of the disasters and therefore realize the importance of prevention and preparedness."

Another feature of the CBDP programme is the revolving fund – set up and managed by the community disaster committee. It may be used to make one-off payments, for example, to villagers who have lost homes or livestock due to disaster. Villagers are encouraged to contribute to it on a monthly basis. Some communities are better at this than others. Of the 23 communities targeted by CBDP activities in Dharan district, for example, 14 have a revolving fund exceeding 54,000 rupees (US$ 725).

The CBDP programme, however, suffers from an acute shortage of resources. As a result, NRCS decided to spread the message of disaster preparedness as widely as possible, targeting communities with training and financial support for just one year each. In some areas following the NRCS's withdrawal, DP activities have not been maintained and local commitment has faded. "Before we could even stand on our feet to realize our own strength as a community, the motivating factor – the CBDP – retracted," said one disgruntled village leader in western Nepal.

Furthermore, to develop the programme across all the rural areas of Nepal would be a huge venture, especially considering that so far the NRCS has only targeted communities able to contribute cash to revolving funds and labour towards preparedness and mitigation projects. More resources will be needed to reach poorer and more inaccessible mountain villages. But exposed communities cannot be expected to shoulder all the responsibility. ■

the 1997-98 El Niño illustrates how risk reduction can get entangled in party politics. Peru's autocratic president, Alberto Fujimori, was described by *The Economist* in May 1998 as "The true artist of El Niño... His government spent US$ 300m in advance (not all in the right places, but at least the ones that looked right at the time); and since El Niño struck he has rushed about frenetically taking personal charge of relief efforts." Peru's civil defence, responsible for national disaster management, was shoved aside by a new group controlled by the president, and the result was organizational confusion. The main beneficiary was

Fujimori himself, whose poll ratings went up from 30 per cent in mid-1997 to 45 per cent a year later.

In many countries, central government has delegated responsibilities, including disaster management, to local government. This has changed the ways in which communities and local NGOs interact with the state to reduce risk. In the Philippines, a national NGO, the Corporate Network for Disaster Response, has been helping local government units to draw up mitigation strategies, and there are signs of growing collaboration between government and civil society in local-level disaster management. When researchers José Luis Rocha and Ian Christoplos looked at disaster preparedness and mitigation in Nicaragua in the aftermath of Hurricane Mitch, they found "creative and surprising alliances" being forged between overstretched municipal authorities and NGOs, despite a tradition of uneasy relations between the two sectors.

However, decentralization can undermine risk reduction efforts. Cash-strapped central governments may simply abdicate their responsibilities, leaving local government and NGOs to take on the task of managing disasters even though they often lack the skills and financial resources to do so. Katrina Allen, who has researched community-based initiatives in the Philippines, identifies another fundamental, but less visible, weakness of decentralization. "There is a danger of depoliticising vulnerability issues," she says, adding that "responsibility for implementation is placed upon those who do not have the jurisdiction or political power to address wider factors and processes which contribute to vulnerability." When this happens, risk reduction is fragmented into a series of small-scale initiatives, focusing on particular hazard events, and artificially separated from broader conditions of vulnerability and ongoing development programmes.

Cuba's success in saving lives through timely evacuation when Hurricane Michelle struck in November 2001 gives us a model of effective government-driven disaster preparedness. This is all the more impressive when one considers that Cuba, although possessing a strong and authoritarian central government, is a poor country. What was the secret of Cuba's success? This is a very complex question to answer. We still do not understand how different types of national or local governance encourage or undermine effective preparedness and mitigation. Ben Wisner, in calling for more systematic study of this neglected but fundamental issue, suggests that good governance in risk reduction might display what he calls a "golden dozen" key features:

- Social cohesion and solidarity (self-help and citizen-based social protection at the neighbourhood level).
- Trust between the authorities and civil society.
- Investments in economic development that explicitly take potential consequences for risk reduction or increase into account.

- Investment in human development.
- Investment in social capital.
- Investment in institutional capital (e.g., capable, accountable and transparent government institutions for mitigating disasters).
- Good coordination, information sharing and cooperation among institutions involved in risk reduction.
- Attention to the most vulnerable groups.
- Attention to lifeline infrastructure.
- An effective risk communication system and institutionalized historical memory of disasters.
- Political commitment to risk reduction.
- Laws, regulations and directives to support all of the above.

Above all, a "minimally trusting relationship" between governments and people at risk is needed. "In the end," says Wisner, "one cannot 'fix' disaster risk with technology alone. It is also a matter of enacting and enforcing laws, building and maintaining institutions that are accountable, and producing an environment of mutual respect and trust between government and the population."

Where this trust between government and people breaks down, the consequences can threaten politicians as well as exposed populations. When the Algerian prime minister and interior minister visited working-class districts of Algiers, where nearly 600 people had been killed by mudslides on 10 November 2001, they were met by residents who were angry because they believed the government had made little effort to protect them: the crowd shouted "Murderers!" and "Long live Osama bin Laden!"

## Innovations offer hope for future

The gradual erosion of the old technocratic, command-and-control approach to risk reduction opens the door for innovative approaches and partnerships. Recent developments may offer considerable potential:

**New concepts and planning tools.** One of the most exciting conceptual advances is the "sustainable livelihoods" perspective (see Box 1.4). Crucially, this views poor people within the external environment where they live and which is responsible for many of the hardships they face. This context comprises all kinds of vulnerability, from long-term population and economic trends to seasonal shifts in food availability, and it specifically includes external shocks such as epidemics .and natural disasters. Placing vulnerability to external shocks at the heart of livelihoods analysis is a big step forward in development thinking. It has already been taken up enthusiastically by several international agencies including CARE and Britain's Department for International Development (DFID).

At field level, some valuable tools exist to assess local people's vulnerability and resilience, and mobilize communities and agencies to take action. The original capacities and vulnerabilities analysis (CVA), which dates back to the 1980s, has been adapted in many places, notably by the Citizens' Disaster Response Network to organize community-based DP in the Philippines. Second-generation models, such as the International Federation's vulnerability and capacity analysis (VCA), are now being applied. To date, these methods have been used mostly by disaster management organizations and often applied to specific hazard threats. Yet they are capable of much broader application, to the multiple vulnerabilities faced by communities, and they can easily be used in long-term development work. Chapter 6 shows how VCA boosted the Palestinian Red Crescent's understanding, capacity and relationships with communities and other disaster responders.

**Innovations in disaster insurance.** The concept of disaster insurance is not new. An insurance policy provides cash payouts following disaster, potentially helping fund the recovery process. Insurance policies can also be made conditional upon implementing certain building and land-use zoning codes, thus acting as a mechanism to enforce risk reduction.

In developed countries there are well-established markets for insurance against a wide range of natural hazards. However, the cost of such insurance fluctuates, depending on the scale of bills incurred by the industry. In 1992, for example, prices leapt three- or fourfold as insurers faced record claims following Hurricane Andrew. This stimulated interest in innovative insurance tools, such as weather index-based insurance. Under such policies, automatic payouts are made within 72 hours of the pre-determined trigger event occurring (e.g., based on earthquake intensities, temperature levels, precipitation over a specified period, wind speed).

There has been recent interest in helping poorer countries gain greater access to international insurance markets at an affordable, relatively stable price. Insurance cover is typically far less extensive in developing countries, with some governments arguing that, in the event of disaster, international assistance will be forthcoming anyway. However, on both a public and private basis there are strong arguments for increasing insurance coverage in developing countries. The World Bank, for example, is supporting a compulsory earthquake insurance scheme established in Turkey following 1999's devastating earthquakes. Insurance is also being tested as a means of protecting local-level savings and credit schemes run by NGOs (see Box 1.5).

**Partnerships with business.** The growing global power of business, compared to that of governments and even inter-governmental institutions, has led to much discussion about the role of the business sector in risk reduction. Some aid agencies have expressed enthusiasm about 'inter-sectoral' partnerships between the commercial, state and non-profit sectors.

The private sector's commercial interests in disaster mitigation – for example, in insurance, engineering and information technology – are well known. But a recent report by the London-based Benfield Greig Hazard Research Centre, which studied businesses' wider involvement in non-profit mitigation work specifically aimed at the public good, concluded that "there is little understanding of what this means in practice and still less of how to go about it". To date, the concept of corporate social responsibility has had little impact in the field of risk reduction. Most of the experience comes from the United States, where such initiatives as the Institute for Business & Home Safety's Showcase State programme and the Federal Emergency Management Agency's local-level Project Impact programme are bringing businesses, local officials and communities together to identify risks and vulnerabilities, raise awareness and plan mitigation measures. There are many challenges in sustaining these initiatives involving businesses – and their failure to address the deeper causes of vulnerability remains a cause for concern – but there is potential here that deserves to be tested more thoroughly.

**A right to safety?** The idea of a right to safety from disasters, to sit alongside other basic human rights, is gaining ground. The idea fits well within the rights-based approaches widely adopted by humanitarian and development agencies over the past few years. It is arguably implicit within other internationally agreed declarations on human rights. The right to adequate food and the responsibility of states to alleviate hunger are already recognized in international agreements.

Safety is difficult to define, since notions of acceptable risk are relative. Decisions about risk and safety may have to be taken where the precise nature, magnitude and extent of a hazard are unclear or disputed. Who is ultimately responsible for ensuring the safety of the public and mitigating hazards? The concept of a right to safety is likely to be challenged by those who fear it will increase their own liability – for instance, governments and the private sector. In any case, risk can only be reduced; it cannot be eliminated.

Despite such difficulties, rights-based thinking could mark a significant step forward in the way we approach risk reduction because it may strengthen lines of accountability and build trust between vulnerable people and those who are supposed to help them. One way ahead could be through an intergovernmental panel to set and monitor international standards. These standards could eventually be linked to the proposed International Disaster Response Law (see Chapter 9, Box 9.1).

## Box 1.4 A livelihoods context for disaster management

On 26 January 2001, a series of earthquakes, peaking at 7.9 on the Richter scale, shattered the Indian state of Gujarat. Officials later put the death toll at over 20,000. Hundreds of thousands of homes were destroyed, and many schools and hospitals collapsed. Up to 15 million people were affected.

While physical, tangible assets such as stronger homes and hospitals are crucial to reduce risks from disasters, there are many less tangible assets which people depend on to recover and survive. For example, in order to benefit from the government's complex compensation scheme, some survivors found that friends in high places proved very useful.

According to an evaluation by the London-based Disasters Emergencies Committee (DEC), one villager said, "We received 2,000 tents for 900 households because we had a prominent politician in the community." Some villagers proved more capable than others in accessing aid for relief and reconstruction. Why? The DEC's evaluation found that "women, lower caste groups and those representing smaller numbers stated they were left out of decision-making in the relief committees and hence were also omitted from relief distributions".

The livelihoods-based approach to disaster reduction tries to unpack these different aspects of vulnerability and capacity. It describes how people, both rich and poor, access the assets they need, how these assets are controlled and how assets are used both to improve livelihoods and to reduce vulnerability to disasters and "shocks" such as ill-health or unemployment. Tangible assets can be both physical (e.g., food, relief, safe hous-ing) and financial (such as income, savings, insurance). However, non-tangible assets are just as important. They include alternative skills, training and disaster awareness (human assets); community organization, self-help and solidarity (social assets); representation in decision-making and the ability to lobby leaders for action (political assets). These non-tangible resources are often ignored by disaster managers, but prove pivotal in sustaining disaster preparedness, mitigation and rehabilitation.

Following the super-cyclone that devastated India's Orissa state in 1999, Britain's Department for International Development (DFID) piloted a livelihoods-based approach to rehabilitation. Financial assets were strengthened through cash-for-work programmes, while cyclone-resistant reconstruction projects enhanced the communities' physical asset base. Significantly, however, non-tangible assets were also developed, such as skills training to improve earning opportunities; raising awareness of vulnerable people's rights; building the capacity of self-help community groups; and strengthening the involvement of the poor in the decision-making process.

The livelihoods-based approach has also been shown to pay dividends in terms of disaster mitigation. In 2000, the worst floods for a century rushed through Mozambique's capital Maputo, rapidly carving out deep ravines which devastated large areas in two of the city's poorest neighbourhoods. The torrents destroyed many houses and the water supply, and threatened to swallow the health centre that served most of the locality.

When the floods hit, an existing livelihoods-based project was being implemented in the same area. This project was focused on reducing poverty by building links between the local residents, municipality, private sector, government, university and NGOs. These links, effectively social and political assets, were instrumental in the setting up of mechanisms, within the municipality, to coordinate development support for poor neighbourhoods.

Significantly, during a recent review, municipality officials, the district administrator and residents said that the relationships built up during the livelihoods project also strengthened their ability to respond to the disaster. Decision-makers now have a better understanding of residents' livelihoods, which in turn has generated more options to choose from in addressing post-disaster infrastructure needs. This could include community-managed maintenance of water systems, ravine repairs and solid waste disposal – measures which would themselves reduce the risks of future disasters.

The livelihoods approach therefore sits on the crossroads between disasters and development. It makes clear that disasters are part of everyday life, and must be overcome if a livelihood is to be sustainable. Within this approach, disaster mitigation is in effect the act of building up tangible and non-tangible assets to reduce vulnerability.

This leads on to another key feature of livelihoods thinking: the need to view vulnerable communities in a holistic rather than a sectoral way. The livelihoods approach sees people as the starting point of all interventions to reduce risk. People's lives are complex and do not fit neatly into the sectoral areas that aid practitioners specialize in. For example, a house is much more than just a shelter – it can be a home, a place of learning, a means of income or an investment. And solidarity among neighbours and their willingness to help in times of disaster is more valuable than the best-drafted preparedness plan. By rooting risk reduction in a developmental context, livelihoods strategies enable disaster managers to take better account of the complex interactions of life that people themselves employ to mitigate, respond to and recover from disaster. Three key priorities have emerged from recent experience:

**Build non-tangible assets.** Improving the skills, self-help and solidarity of households and communities will prove as important in the face of disaster as investing in physical and financial defences.

**Strengthen everyday lives.** Preparing for major disasters is only part of risk reduction. Smaller, ongoing disasters can, over a period of time, take a heavier toll than the big one-off disasters. So strengthening everyday lives by investing in human, social and political assets will help reduce the risks posed by a whole range of hazards, large and small.

**Listen to local priorities.** The livelihoods approach puts vulnerable people and their priorities at the centre of aid strategies. Despite much rhetoric, this often doesn't happen. As the DEC's evaluators in Gujarat discovered: "People constantly emphasised the need to restore livelihoods rather than receive relief and expressed some frustration that outsiders did not listen to them on this point. They wanted to receive cloth and make their own clothes rather than receive clothing but no one took any notice." ■

# Targets for risk reduction

The threats that natural hazards pose to human society and sustainable development are undoubtedly massive. The scale, extent and complexity of disasters and vulnerability present enormous challenges. Our understanding remains incomplete; our organizational capacity, financial resources and tools are still woefully inadequate.

Yet we know that disaster mitigation and preparedness pay – in human, economic and environmental terms. There are success stories to guide us, from across the world. Innovations are opening up opportunities to make more progress in the near future.

How, then, can we ensure that we really advance the thinking and practice of risk reduction over the next few years? Here are three suggestions that, if followed, could radically reform the way we deal with risk and vulnerability.

**Relocate disasters within the wider context of risk reduction.** Disasters are, after all, just one aspect of risk, and risk management should be everyone's concern. Even though many people's understanding of specific risks is imperfect, risk as a broad concept is commonly understood, and risk assessment now forms an essential element in many planning processes, from business to engineering to social development.

Redefining disaster mitigation and preparedness as aspects of risk reduction could break down the many cultural, institutional and methodological barriers separating relief and development professionals. For too long, disaster management has been viewed, and organized, as a separate sector. This separation has been intensified by artificial divisions within the sector, between those who approach mitigation and preparedness from the direction of humanitarian relief and those who approach it from a developmental perspective.

Risk reduction terminology can be applied across the relief-development spectrum. It can be applied to all types of risk reduction activity, from early warning systems, stockpiling relief materials and preparedness for response through to advocacy for greater social and economic equity to reduce vulnerability. It can be applied at all levels, from the local to the global, and by every kind of institution, from the village community to multilateral and inter-governmental organizations. Risk reduction is not exclusive to the big disasters that preoccupy aid agencies, but can be applied to the numerous smaller hazard events that undermine vulnerable households.

**Long-term partnerships based on good governance.** Risk reduction is a long-term process, not a one-off intervention. Viewing disasters in this way steers us away from

## Box 1.5 Insuring micro-finance institutions against disaster

Micro-finance institutions (MFIs) are beginning to talk about disaster insurance. MFIs provide financial services to the poor, extending credit and providing savings facilities. The loans they provide are typically very small, are mainly intended for productive purposes, do not require conventional forms of collateral and are extended on a non-profit-making basis. Many of their clients would not be able to obtain such loans from the commercial banking sector. MFIs thus provide a very important service, helping the poor to invest in new productive activities that will increase their livelihoods, and enabling them to access the funds they need to recover from seasonal shocks, such as flooding.

Reflecting their client base, MFIs themselves are highly vulnerable to natural hazards. As Warren Brown and Geetha Nagararajan observed in the context of the 1998 floods in Bangladesh, they can face temporary liquidity difficulties as they simultaneously try to support clients through difficult periods whilst also experiencing a decline in flows of debt repayments as people are temporarily unable to meet their dues. Not only do natural hazards threaten the survival of borrowers, but the very assets purchased with previous loans (for example, agricultural tools or chickens) may have been lost, threatening their recovery. In Bangladesh, for example, the Grameen Bank, the original pioneer of micro-credit operations, reported that around 1.2 million of its 2.3 million members were affected by the 1998 floods, of which 0.8 million were seriously affected.

However, it is also recognized that it is important not to encourage a culture of default. Hence, in the longer term, borrowers are often expected to honour their loans, even when the activities funded through them have been destroyed.

Some MFIs are therefore beginning to explore options for disaster insurance, both to protect themselves and to enable them to respond to additional disaster-related needs for their clients. This interest has been partly motivated by the discovery that the poor may use loans as a de facto insurance policy, to pay for consumption and survival needs, or to replace basic means of production after a disaster. To date, those MFIs that have established schemes have basically opted for self insurance, setting some resources aside into a calamity fund for use in the event of an emergency. In the event of a disaster seriously affecting a significant proportion of clients, however, such funds would be grossly inadequate. The alternative, placing the risk externally, would create additional overheads, making the cost of credit itself more expensive.

A major challenge ahead in the fight against poverty and vulnerability is thus to devise some way of supporting MFIs, either individually or as part of a group, in establishing some form of viable but affordable insurance. ∎

the "technical fix" towards broader strategies that address human vulnerability, strategies that are more people-centred and less hazard-centred. Disasters affect *people*, after all, and the evidence shows that they are affecting more and more people every year.

Disasters are unsolved problems of development, which means they are therefore problems of governance, in its broadest sense. Good governance needs to be placed at the heart of risk management. Disasters are complex problems, requiring complex solutions that draw on many different skills and capacities. Instead of top-down disaster management based on fixed-term projects, we need long-term partnerships for risk reduction involving multiple stakeholders, drawing on their different capacities and respecting their different needs.

Such partnerships cannot be imposed: they must be negotiated, and built on trust and confidence. Every partner must have a voice at the negotiating table, especially the most vulnerable communities, which have much to contribute. Creating trust will be difficult in many countries, given the history of mutual suspicion between governments and civil society, and between business and the public sector. But, as we have seen, alliances can be forged in unlikely situations, and each successful partnership is a building block for others.

Good governance – at all levels, and in all types of group and organization – depends on two fundamental principles: participation and accountability. All partners should participate in making decisions about the processes and initiatives that affect them. Only in this way can we identify needs, capacities and priorities accurately, define problems correctly, and design and implement appropriate risk reduction measures. Accountability means finding mechanisms by which partners' performances can be judged and they can be held responsible for their actions. All too often, disaster management professionals recognize their accountability to bosses, boards of management, donors and governments, but fail to recognize that they should be accountable above all to the people they claim to be helping: disaster victims and vulnerable communities (see Chapter 7). This balance has to change.

**Targets for risk reduction.** Fine sentiments need to be turned into action. Organizations working for disaster mitigation and preparedness could promote trust, accountability and innovation through one simple action: setting targets for risk reduction. Setting targets forces agencies and governments to square up to this issue: there can be no hiding behind well-meaning generalizations and easily agreed principles. Targets provide a benchmark for judging their commitment. Targets can be set by everyone, at every level.

The idea may seem simplistic, even impractical. But International Development Goals have been adopted by the World Bank, the International Monetary Fund and the Organisation for Economic Co-operation and Development's donor nations, and endorsed by the UN General Assembly. These goals are ambitious, setting out 21 indicators to measure progress by 2015: for example, halving the proportion of people living in extreme poverty in developing countries (compared to 1990 levels);

universal primary education in all countries; reversing the spread of HIV/AIDS; a two-thirds reduction in death rates among infants and children under five in developing countries; and implementing a national strategy for sustainable development in every country.

If targets can be set for sustainable development, then why not for risk reduction? National governments could, for example, set targets for reducing the numbers of people killed and directly affected by disasters, based on annual averages over rolling ten-year periods. They could set targets for designing and implementing national disaster management plans – which was supposed to have happened during the IDNDR but didn't in many cases. Local governments, NGOs, and communities could set targets for designing and implementing mitigation and preparedness plans, training emergency response teams, establishing early warning and evacuation systems, protecting lifeline infrastructure (such as hospitals), and reversing environmental degradation such as deforestation on unstable hillsides. Businesses could commit themselves to protect their employees, suppliers and clients.

Because developing-country governments and civil society lack resources for some of these measures, donor governments and agencies could set targets for allocation of resources to risk reduction. This could mean devoting a percentage of both official development assistance and emergency relief to disaster mitigation and preparedness initiatives.

Above all, communities, agencies and governments alike need to act now with a sense of urgency to prevent the unnecessary suffering of hundreds of millions of people every year. On average, more than 1,000 people lose their lives to natural disasters every week. Direct costs of these disasters amount to well over a billion dollars each week. Only coherent, long-term and well-resourced initiatives will make any impact on reducing these unacceptable losses.

*Principal contributors to Chapter 1 were John Twigg, an Honorary Research Fellow at the Benfield Greig Hazard Research Centre, University College London, and Charlotte Benson, an economist who has ten years' experience in research on the economic aspects of natural disasters. Jonathan Walter, editor of the* World Disasters Report, *John Bales (International Federation disaster preparedness delegate, Bangladesh) and Gawher Nayeem Wahra (editor,* Bangladesh Disaster Year Report*) contributed to Box 1.1; John Twigg and Madhavi Ariyabandu (Intermediate Technology Development Group, South Asia) to Box 1.2; Devinder Sharma, a journalist based in India, to Box 1.3; Andrea Rodericks and David Sanderson of CARE International UK to Box 1.4; and Charlotte Benson to Box 1.5.*

## Sources and further information

Blaikie, Piers et al. 1994, *At Risk: Natural Hazards, People's Vulnerability, and Disasters.* London: Routledge, 1994.

Brown, Warren and Nagarajan, Geetha. "Bangladeshi experience in adapting financial services to cope with floods: implications for the microcredit industry" in *Microenterprises Best Practice. Mimeo.* Bethesda, MA: Development Alternatives, 2000.

"Emerging Perspectives on Disaster Mitigation and Preparedness" in *Disasters,* Vol. 25, no. 3, September 2001.

Gellert, Gisella. *Algunas lecturas de riesgo y vulnerabilidad en Guatemala, utilizando la herramienta Desinventar.* Guatemala: FLACSO, 1999.

Peppiatt, David. *De-naturalising disasters: Reflections on the changing discourse of natural disasters.* British Red Cross Society Working Paper, 2001 (unpublished).

Twigg, John. "The Age of Accountability? Future community involvement in disaster reduction" in *Australian Journal of Emergency Management,* Vol. 14, no. 4, pp. 51-58.

## Web sites

Benfield Greig Hazard Research Centre **http://www.bghrc.com**
DesInventar disaster database for Latin America **http://www.desinventar.org**
EM-DAT international disaster database **http://www.cred.be/emdat/intro.html**
International Federation **http://www.ifrc.org**
NGO Initiatives in Risk reduction (set of 19 case studies on British Red Cross
    web site) **http://www.redcross.org.uk/riskreduction**
Organization of American States **http://www.oas.org/en/cdmp/**
ProVention Consortium **http://www.proventionconsortium.org**
Radix (Radical Interpretations of Disaster)
    **http://www.anglia.ac.uk/geography/radix**

Section One

**Focus on
reducing
risk**

# Disaster preparedness — a priority for Latin America

It was the perfect disaster. Hurricane Michelle ripped through Cuba in early November 2001. It was the most powerful hurricane to hit the country since 1944, with winds reaching 225 kilometres per hour. But just five people died. Successful civil defence planning, augmented by the local Red Cross, ensured that some 700,000 people were evacuated to government-run emergency shelters in the hours before the hurricane struck. Search and rescue and emergency health-care plans swung into action in the hours afterwards. In Havana, electricity was turned off to avoid deaths from electrocution when power lines came down, and water supplies were turned off to avoid contamination with sewage. Cuba's population was advised in advance to store water and clear debris from streets that might cause damage. As a United Nations (UN) interagency mission reported five days afterwards, the government's "high degree of disaster preparedness...was decisive in the prevention of major loss of life".

The post-hurricane misery was real enough, of course. Homes, infrastructure and crops were destroyed on a large scale. "Many people have lost everything," said Cuban Red Cross relief coordinator Virginia Huergo. But with most human life secure, domestic and international aid could immediately concentrate on relief and rehabilitation. The emergency phase of international response to such a disaster, which generally lasts for several weeks, was all but eliminated. The UN mission reported that, less than a week after the disaster, local authorities were already delivering construction materials to families whose houses were damaged.

Cuba's response to Michelle, says Nidya Quiroz, the UN Children's Fund (UNICEF) regional emergency adviser in Panama City, was a textbook case of successful disaster preparedness. "This is how to do it. Cuba is a poor country with many problems, but they are teaching the rest of us – even in rich countries – how to respond." Of course it helped to have a centralized government in which "Fidel controls everything," she agrees. But the way that Cuba deploys its expertise at the local level to help poor communities in particular has lessons for everyone. "The doctors and nurses all knew their role. They had teams for every activity and stockpiles of medicines."

The contrast between events in Cuba last year and comparable "natural" disasters in the region a couple of years before – Hurricanes Mitch and Georges in 1998 and the floods in Venezuela in 1999 – is a subject of anguished debate among governments and aid agencies in the region. Mitch, in some respects a lesser hurricane than

Photo opposite page:
More investment in disaster preparedness is urgently needed to ensure all exposed communities are less vulnerable to disasters. In Honduras, in the aftermath of Hurricane Mitch, people are rebuilding houses designed to better withstand natural disaster.

Yoshi Shimizu/ International Federation, Honduras 1999.

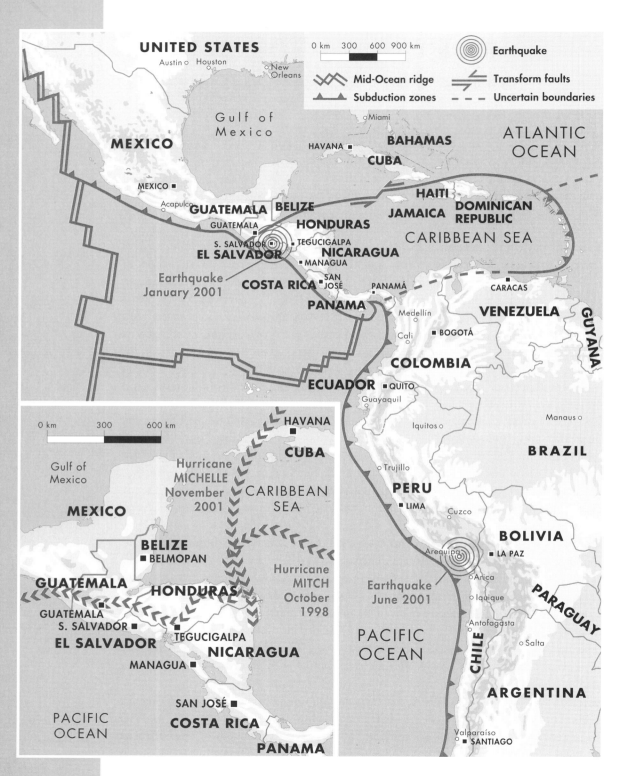

UNITED STATES

Austin ○   Houston ○

○ New Orleans

0 km    300   600  900 km

Earthquake

Mid-Ocean ridge

Transform faults

Subduction zones

Uncertain boundaries

Gulf of Mexico

○ Miami

MEXICO

HAVANA ▪

BAHAMAS

CUBA

ATLANTIC OCEAN

MEXICO ▪

Acapulco ○

GUATEMALA ▪ BELIZE

GUATEMALA ▪

HONDURAS

HAITI

JAMAICA

DOMINICAN REPUBLIC

S. SALVADOR ◉   ▪ TEGUCIGALPA

EL SALVADOR

▪ MANAGUA

NICARAGUA

CARIBBEAN SEA

Earthquake January 2001

COSTA RICA

SAN JOSÉ ▪

PANAMÁ ▪

CARACAS ▪

PANAMA

Medellín ○

VENEZUELA

GUYANA

Cali ○   ▪ BOGOTÁ

COLOMBIA

ECUADOR   ▪ QUITO

Guayaquil ○

Manaus ○

Iquitos ○

BRAZIL

○ Trujillo

PERU

▪ LIMA

Cuzco ○

BOLIVIA

Arequipa ◉   ▪ LA PAZ

Earthquake June 2001

○ Arica

○ Iquique

PARAGUAY

PACIFIC OCEAN

Antofagasta ○

○ Salta

CHILE

ARGENTINA

Valparaíso ○   ▪ SANTIAGO

---

0 km    300    600 km

Gulf of Mexico

HAVANA ▪

CUBA

Hurricane MICHELLE November 2001

CARIBBEAN SEA

MEXICO

BELIZE

▪ BELMOPAN

Hurricane MITCH October 1998

GUATEMALA

HONDURAS

GUATEMALA ▪

S. SALVADOR ▪   ▪ TEGUCIGALPA

EL SALVADOR

NICARAGUA

MANAGUA ▪

SAN JOSÉ ▪

COSTA RICA

PACIFIC OCEAN

PANAMA

Michelle, killed up to 20,000 people and, in the words of the Honduran prime minister, put the country's economic development back 20 years.

Some fear that without urgent action to improve disaster preparedness, the region could become caught in a spiral of natural disasters that wreck development and feed a growing vulnerability to each succeeding disaster. But where are resources for disaster preparedness best invested – at local, national or regional level? Others argue that stronger economic development will itself reduce the risk of disasters. That may be true in the very long term. But, given the annual recurrence of natural disasters, and with El Niño's extreme weather conditions returning to the region's Pacific coast every five to seven years, the case for focusing on disaster preparedness has never been greater.

## Increasing vulnerability

Latin America is a crucible for a full range of disasters, from the tectonic to the climatic. According to François Grunewald, who heads a think-tank that advises the French government on crisis management, Central America in particular is "one of the world's most geo-dynamic regions, marked by recurrent seismic and volcanic activity, as well as hurricanes, forest fires and drought". El Niño, though now recognized as a near-global climatic phenomenon, was first identified and has some of its most intense consequences in the region.

There is nothing new about the region's exposure to natural disasters. San Salvador, capital of El Salvador, has been seriously damaged by earthquakes 14 times in the past three centuries. Managua, capital of Nicaragua, is not far behind. Between 1960 and 1988, the Office of US Foreign Disaster Assistance logged 64 natural disasters in the seven countries of Central America alone. But many countries in the region – and particularly their marginalized, poor communities – are becoming increasingly vulnerable to the destructive forces of nature through poor construction and siting of buildings and environmental degradation.

Says Grunewald: "Uncontrolled urban sprawl and speculative land markets have pushed many marginal settlements into high-risk areas such as river canyons and flood-prone coastal zones. The expansion of the agricultural frontier into more fragile ecosystems – eliminating stabilizing forest cover from steeper and unstable terrain – has increased the frequency of flash floods, mudflows and landslides." Such factors clearly played a major role in the high death tolls in Honduras and Nicaragua during Mitch and the floods of Venezuela in 1999.

Tens of thousands of San Salvador's citizens live in ravines and on steep slopes, despite the danger of landslips from the city's extreme seismic vulnerability. During the series of earthquakes that shook El Salvador in early 2001, some 700 of the 1,100 who died were buried when a landslide engulfed 500 houses in the suburb of Las Colinas in Santa Tecla, just outside San Salvador. This, says geographer Ben Wisner of Oberlin College, Ohio, "was not an 'Act of God'. A group of residents and environmental groups were in court only a year before to stop development on that slope and the ridge above" because of the evident risk.

But besides the physical vulnerability of communities, there is an additional factor: the failure of governments and aid agencies to prepare communities, and sometimes themselves, to cope with disaster. An absence of simple evacuation procedures may have been responsible for a large proportion of the deaths in Tegucigalpa, the Honduran capital, when Hurricane Mitch struck, says Red Cross delegate Jan Gelfand. "Half the deaths happened because people went back to their houses to collect their families. If there had been an evacuation procedure so people knew where to meet their families in a safe place, most of those lives could have been saved." Nor were there procedures for efficient search and rescue and post-disaster evacuation, or stockpiles of key relief supplies like food, blankets and tents. The contrast with the Cuban experience during Michelle could hardly be greater, and goes a long way to explaining their hugely different death tolls.

The reasons behind this vulnerability to disaster lie partly in the increasing fragility of government agencies, including civil defence, during a time of economic recession and international pressure to privatize central services. But there is also a simpler reason than economic hardship: a failure to learn lessons from past disasters and engage in cheap but effective disaster preparedness. Some argue that international aid agencies, as well as national and local governments, have failed in this respect – they have lost sight of the real difference that disaster preparedness can make.

## "Physician, heal thyself"

The nature of the current debate on how to prepare for and respond to disaster in the region has been deeply coloured by the three disasters Mitch, Georges and the Venezuelan floods. Numerous international agencies conducted post-mortems into the handling of these disasters by both national governments and their own staff, often with withering results. From these reviews, a picture emerged of disorganized and poverty-stricken national civil defence agencies augmented by proud but hopelessly under-resourced relief agencies, including National Red Cross Societies, overwhelmed by an uncoordinated influx of international aid, much of it brought by people who could not even speak Spanish.

As help flooded in during Mitch, for instance, money, resources and expertise were wasted. "People did their own thing; it was chaos," says Gelfand, who was in

Honduras during Mitch. It was not clear who was, or should be, in charge. Was it the locals with their experience of the country, or the foreigners with their skills in handling other disasters?

A review team assembled after these disasters by the International Federation of Red Cross and Red Crescent Societies under Douglas Lindores, former secretary general of the Canadian Red Cross, reached some hard-hitting conclusions, not least about the Red Cross itself. While Lindores's team praised National Societies' quick response to the disasters, and their capacity to inspire spontaneous "voluntary service", they also accused the International Federation of letting down the often heroic work of these volunteers through "systematic weaknesses", including poor coordination between members of the Movement.

The review team charged National Societies and the International Federation's secretariat with being "not adequately prepared" for their primary remit – "to respond in a timely and effective manner to disasters". The team added that a lack of focus had "diluted disaster preparedness efforts and resources". While they recognized the importance of community-based disaster preparedness, the reviewers suggested that National Societies themselves have a "fundamental obligation" to ensure that they too are "appropriately prepared".

In disaster-prone Central and South America, failing to prioritize disaster preparedness and risk reduction will continue to have catastrophic consequences.

Cecilia Goin/ International Federation, Peru 2001.

# Regional support for local preparedness

The International Federation, perhaps more than some other aid organizations, has confronted the crisis of disaster preparedness and response, post-Mitch. In particular it has created a new regional body, PADRU (the Pan-American Disaster Response Unit), to beef up its disaster preparedness and response capability and help to define and develop the capabilities of National Societies. Iain Logan is its head of disaster management and coordination – he says that the Lindores report provided important impetus to the setting up of PADRU.

PADRU, which could become a model for Red Cross operations in other regions of the globe, is based in Panama, where good communications, security and a free trade zone mean it can procure and distribute goods quickly. In late 2001, its Panama warehouse contained water and sanitation equipment, blankets, satellite telephones and a range of relief equipment supplies such as medical kits and plastic sheeting, some of it recovered from the Mitch and Venezuela flood operations, and ready to be shipped anywhere in Latin America within 24 hours of a request from a Red Cross society. "We hold things that are difficult to get quickly in an emergency," says PADRU's logistics manager, Jon Carver.

PADRU's remit includes:
- encouraging risk mapping and early warning systems;
- stockpiling relief supplies;
- arranging standing "pre-contracts" with suppliers;
- bolstering local and national capacities to organize teams of volunteers and community brigades; and
- establishing emergency response units.

Of these, says Logan, the first priority is strengthening the capacities of National Societies to prepare for and respond to disasters. This work is well under way: PADRU now trains national intervention teams to International Federation standards. A database is being built up of staff, trained in a range of disciplines, who are able to respond at 24 hours' notice. PADRU-trained national staff have worked alongside international Red Cross teams in every disaster in the region since Mitch, says Logan. By training them to respond as part of national and regional intervention teams, Logan hopes that PADRU can be an instrument for the decentralization, rather than the centralization, of relief provision. And for saving more lives.

"We don't expect National Societies ever to have to deal with big disasters alone, but they can be better trained, equipped and more coordinated," says Logan. "The response to the earthquakes in El Salvador and Peru, and Hurricanes Iris and Michelle in the Caribbean, in 2001 showed significant improvements since Mitch. In El

Salvador, PADRU people and resources were on the ground within eight hours, having drawn up a plan of action with the National Society. That quicker response certainly saved lives." And in the Peruvian quake of 2001, says Logan, PADRU combined airlifts from Panama with the activation of pre-contracts with local suppliers in Peru.

There were hiccups in Peru. "At first we didn't know if PADRU came to direct or support us," says Clotilde Villena, president of the Arequipa branch of the Peruvian Red Cross. "For five days it directed, but in the end it provided technical support, which worked much better." But as a result of PADRU's intervention, regionally held stockpiles of supplies reached the area days earlier than would otherwise have been the case. According to Carver, one Peruvian supplier who had committed to provide 10,000 blankets could not meet the order. After a three-day delay, the decision was taken to bring in blankets from outside. "People were freezing; we couldn't wait," says Carver.

Carver is adamant that his job is not to circumvent National Societies, but to augment and strengthen them. A key task of PADRU, he says, is to help source relief supplies locally: "We are now working to develop agreements and pre-contracts with national and regional suppliers which will allow us to deliver relief supplies in a very short time." Edgardo Calderon, the president of the Peruvian Red Cross, emphasizes that PADRU's role must be one of "giving assistance and advice previous to an intervention; building a relationship to know the capacities and necessities of the National Red Cross, with the ultimate goal of helping just in those weak points."

A critical question for planners of this new regional strategy is how much aid should be stockpiled and where. Villena hopes in future to establish a local stockpile of relief materials in Arequipa. "If we had our own warehouse we would be able to respond within 24 hours," she says. Similarly the provincial civil defence chief, Carlos Nacarino Rodriguez, is talking of establishing local stores, holding clothing, tents, blankets and food at several small towns in the mountains. "We are all aware we have not been sufficiently prepared," he says. "With roads broken, some villages could only be reached on donkey or with a ten-hour walk."

At PADRU they are less sure about the wisdom of this approach. Such stocks are expensive and difficult to manage. Critical questions arise: Who owns them? Who has the right to requisition stockpiles for a disaster outside the area for which they were originally created? "It would make more sense to have regionally available stocks held under clear understanding of to whom they belong and who can authorize their use," says Carver. Donors are also attracted by the transparent accountability which regional control of their donations provides. "It would be ideal to have such logistics in each country," says Edgardo Bartomioli, a Lima-based delegate of the German Red

Cross. "But you can see the advantage of a regional facility in Panama. It has good results. It can get supplies here in 24 hours."

This approach to stockpiles exemplifies PADRU's attitude to regionally-based disaster preparedness. It does not want to centralize unnecessarily – indeed often the priority will be to decentralize. But the aim of the regional unit, it believes, is to identify how best to ensure rapid, effective delivery of relief aid while building capacity of local-level disaster preparedness and response. Some resources – such as search-and-rescue personnel – need to be locally based to be effective, while others – such as strategic stockpiles of relief aid – may be better sourced regionally or through pre-contracts with national suppliers. Some skills, such as evacuation procedures, need to be embedded within communities themselves; others, such as management expertise in handling massive international relief, are best centralized.

But many questions remain. Crucially, exactly how much difference can disaster preparedness make to saving lives? Is it right to suggest that economic development is the best form of preparation to withstand natural disasters? This, after all, is the view, in their different ways, of both the neo-liberals of the right, who want unfettered market-based economic growth, and the radical social reformers of the left. Or is there a more risk- and people-centred agenda that both sides, in pursuit of their macro socio-economic agendas, have tended to ignore?

## Reduce disaster risk – not just poverty

One of the key ideas in this debate is the notion of the "class quake". The phrase was born in Latin America to describe a seismic event in Guatemala in 1976 that killed 22,000 almost exclusively poor people in the rural highlands and slums of Guatemala City. It encapsulates the view, widely held today that, as Ann Varley of University College London has put it, "Disasters are just big versions of everyday hazards faced by poor people." The argument is that vulnerability to natural disasters is overwhelmingly a function of poverty, inequality and social exclusion, and therefore that development is the best form of disaster preparedness. The director general of the Panama Red Cross, José Beliz, goes so far as to say that "we need to prepare more for day-to-day social disaster and less for occasional natural disasters".

There is plenty of evidence that the poor are often the most vulnerable, of course. In the months after Mitch, residents of the houses on the steep hillsides in Tegucigalpa that had collapsed so catastrophically were rebuilding as if nothing had happened. "One hillside is today full of people and slipping at a rate of a centimetre a month. The authorities have tried and tried to get people to move, but they won't go," says Carver. It is hard to believe that people would have stayed and rebuilt if they had an alternative.

Such observations have had an important effect on the way many relief agencies see their role. Dilma Davila, head of the projects office of the Peruvian Red Cross in Lima, sees development as her prime task. "Before about 1997, we were focused on handling emergencies, with first aid and so on. Our aim now is to give power to the people, by providing the most vulnerable with tools to help them develop."

But others are not so sure that development will of itself reduce vulnerability. They feel the pendulum may have swung too far. It is true, they say, that shanty towns built on or below steep, deforested slopes, or on flood plains, have suffered greatly from hurricanes and floods in the region in recent years. But during Peru's earthquake last year, some shanty dwellers survived where richer neighbours perished (see Box 2.1). And the 700 victims of the landslip at Las Colinas near San Salvador in the 2001 quake were mostly middle class.

Clearly forces other than poverty are exposing people to disasters. Lack of land zoning regulations may allow developers to build in high-risk areas. Corruption may allow them to ignore existing building codes and regulations. Ignorance may mean that

## Box 2.1 Traditional homes prove safer

Sometimes the poor survive best. Take the town of Punta de Bombon in southern Peru. It suffered badly in the 2001 earthquake. Whole street blocks disintegrated and five months on there were few signs of repair. But perched on the hillside above the town were extensive shanty settlements barely troubled by the quake. The houses, made of flexible reeds and easily reassembled sheets of corrugated metal, were either untouched or quickly repaired. The poorest, most marginalized families in the area survived last year's quake relatively untouched, says the Red Cross's rehabilitation coordinator Freddy Gonzalez.

Shanties, it seems, can sometimes be the safest place in a quake. The light materials that comprise most self-built homes of the poor are much less dangerous than concrete or masonry. Many people tell stories like that of Panamanian Red Cross volunteer Leonar Arboleda. When her family was caught in a hurricane in Nicaragua in 1986, she remembers, "My aunt and cousin left their shack to find shelter in a big Christian church. But the church collapsed and killed them, while the shack was unscathed."

Many argue that the traditional building materials and designs of the pre-Columbian Andes, which are often still used by the rural poor, are much safer because, unlike modern construction methods, they were designed with quakes in mind. Inca buildings, according to Tony Oliver-Smith, an anthropologist at the University of Florida, had thatched roofs, little masonry, were single-storeyed and avoided long overhead beams. And the Incas kept settlements small, well-spaced and clear of valley floors. Modern planners and developers could benefit by applying traditional building techniques. ∎

people move into sub-standard buildings blind to the risks. And if nobody has bothered to map out where the high-risk areas are, no amount of social planning or wealth will provide protection. Rich and poor died together when lava belched from a Colombian mountainside in 1985 and obliterated a city of 20,000 people. Riches would not have saved one of them; a decent prediction of the eruption and an evacuation procedure could have saved them all.

The development agenda has often submerged genuine and important debates about managing risk, says José Luis Rocha of the Central American University in Managua. In Nicaragua since Mitch, "disaster prevention and mitigation has by and large been overshadowed by the national debate over different development models. Shockingly, many reports and recommendations on the requirements for rehabilitation and future development totally ignore the impact of natural disasters on these alternative development scenarios, and on the impact of these development scenarios on the ability of vulnerable populations to withstand shocks to their livelihoods."

According to Ben Wisner, the government in El Salvador, which is "sold on privatization", has been "resisting popular demands since Mitch to reform the whole system of emergency response". Yet given how disaster-prone Latin America is as a region, a failure to prioritize disaster preparedness and risk reduction will continue to have catastrophic consequences.

Wisner points out that the regional Pan American Health Organization (PAHO) "has been a leader, especially in providing very straightforward design and technical assistance for protection of hospitals and clinics. The trouble is that national governments sometimes only pay lip service and don't follow the advice." Where governments fail to take the lead, can regional humanitarian and development organizations seize the initiative? According to Iain Logan, "the Red Cross, at all levels, has a significant advocacy role and opportunity to advise government in risk mitigation, planning and the development of national disaster strategies."

One problem for regional organizations, says Wisner, is the political requirement to work through "national centres of power". But "they should reach out to civil societies and municipalities" as well. He stresses that municipalities are places "where the potential for rapid and meaningful change can really take place". Clearly, the national level is the only level at which certain changes can be achieved – such as generating the political and legal will necessary to enact and enforce better building codes. But, he argues, "at community level, one can see evolving a strong demand for national action as well as a neighbourhood 'culture of prevention'". This is the "top priority for disaster preparedness in Latin America – to establish civilian emergency management systems in cities [that] draw on civilian groups of various sorts in making contingency plans".

# Community preparedness pays

So, while the debate over models of development will continue for years, there is consensus over the value of community and municipal disaster preparedness. On the national scale, we have already seen that a poor and hard-pressed country such as Cuba can prepare itself effectively to withstand disaster. But even individual communities can do the same. During Mitch, those few Nicaraguan communities that had had prior experience during the recent civil conflict in self-help organization "proved to be very effective in evacuating the population and distributing aid", says Rocha.

And, in poor communities around Arequipa, Peru's second city, the timely establishment by the local Red Cross of 15 emergency brigades with basic training in evacuation and first aid and with links to local authorities and civil defence structures directly saved lives during 2001's earthquake. The night before the quake, the brigade in the slum of Pampas Polanco rehearsed evacuation procedures and learned how to pitch a tent as shelter in the aftermath of disaster. Five months later, amid the wreckage, brigade members still joked about how the Red Cross team must have known a quake was coming.

Moises Rosales – dentist, long-time Red Cross volunteer and local director of the project – says that the 30,000 people under the umbrella of the Arequipa brigades fared much better in the minutes and hours after the quake. Elsewhere across southern Peru hundreds died as they returned home to collect loved ones or cowered in panic in buildings that they believed to be safe. But in Pampas Polanco, "they knew what to do and where to go as their houses collapsed. They didn't panic. They worked together as neighbours, getting everybody to open spaces. The brigades definitely saved lives here."

Brigade member Beatrice Larico told of how her grandfather survived because he escaped the shack in her yard where he lived before it collapsed. And even in the rubble of their homes – built in an old quarry whose walls partly collapsed onto the community – the brigade women proudly showed the Red Cross first-aid box they keep stocked for the next emergency. Field-tested in a real disaster, the brigades are now to be replicated, with European Union funding, more widely across southern Peru's earthquake zone. The lesson is that even the poorest communities can be made safer from natural disasters.

# Early warning saves lives

Earthquakes are rarely forecast, but extreme weather can be, with great success. However, timely hurricane forecasts, for instance, are of little use unless communities

## Box 2.2 Community-operated early warning in Guatemala

Spanning an area of 900 square kilometres, the Coyolate River basin encompasses both volcanic highlands and fertile coastal plains. The upper basin is planted with coffee, and the middle and lower basins with sugar cane, African palm and bananas. The river has flooded several rural towns in the flood plains on a yearly basis, prompting Guatemala's national emergency council to seek resources for flood-plain risk reduction. The resulting project began in 1997 with support from the Swedish International Development Agency (SIDA) and involved designing and implementing a community-operated early warning system, along with other risk management measures.

Early warning is the first line of defence against natural hazards, especially floods. The early warning system was designed to involve at-risk communities in all aspects of its operation. Plastic rain gauges to monitor rainfall and simple electronic instrumentation to measure river levels were installed throughout the basin. The project established local emergency committees, selected community volunteers, and trained them in early warning through simulation exercises. The project also helped communities to develop hazard maps, establish emergency plans, create specific committees to deal with search and rescue, shelter management and security, and maintain and extend dykes to prevent floods.

Community volunteers now use these simple instruments to transmit information on rain and river levels, via a solar-powered radio network, to a local forecasting centre. The centre, staffed by members of the local emergency committee, is then able to forecast floods two to three hours in advance and initiate emergency preparations. If serious flooding is imminent, the committee may issue alerts using sirens, bells or a public-address system.

Since its inception, the Coyolate early warning system has benefited more than 5,000 inhabitants in around 100 flood-prone communities. It proved its strength during Hurricane Mitch in 1998, when the flood information it provided to authorities helped save dozens of lives. The cost of setting up such a system starts at around US$ 50,000, depending on the scale of the project. That's about a quarter of the price of the cheapest telemetric system, which measures rainfall and river levels using sophisticated stand-alone instruments.

All community-operated early warning systems are based on three pillars: voluntary community observers with the necessary tools, training and continued institutional support; simple, practical instruments to measure rainfall and river levels, supported by a radio network and sustained by the national emergency management institution; and recognition by that institution that risk management and disaster response can initially be handled at a local level.

The key point with all community-operated early warning systems is to ensure they are implemented with the full support of the national emergency agency, which must continue to support the system once the set-up phase is finished.

Coyolate's success has led to similar systems being developed throughout Guatemala and Central America. There are now more than 20 community-operated early warning systems in watersheds throughout the region, most of which have been implemented since Hurricane Mitch. ■

are warned and know what to do. In Cuba, effective civil defence planning and dissemination of information through state-run media allowed the country's evacuation procedures to be activated in the hours before Michelle struck. Few such systems were in place three years before in Central America when Mitch hit, but some communities had devised their own systems.

Jorge Ayala of the Centre for the Prevention of Natural Disasters (CEPREDENAC), a regional organization based in Panama, cites a flood-protection project on the Coyolate River in Guatemala (see Box 2.2). Towns along the river got together in the mid-1990s to map flood hazard zones, build shelters and monitor river levels. The first alarm is triggered by rainfall gauges in the mountain headwaters. It alerts communities in the upper reaches of the river to check river flows. Then, as flows reach danger point, communities downstream are alerted to head for the shelters.

Soon after the Coyolate system was established, Hurricane Mitch came along. The project, which cost little to install, probably saved dozens of lives. While almost 300 people died in floods along other rivers in Guatemala, on Coyolate "there was no loss of life...during Mitch, because the people downstream were successfully evacuated before the floods hit," says Ayala. As a result the charity CARE USA has funded similar community schemes along several more flood-prone rivers in Guatemala.

It sounds simple, but such basic monitoring and communication systems for the natural environment remain rare in Latin America, because of the level of social and political organization needed to establish and maintain them. Ayala makes the point that these community-based early warning systems must be recognized and supported by the national emergency agency or civil defence. Without this integration of local and national levels, community-based disaster preparedness will prove less effective and harder to sustain.

Predicting El Niño, the climatic flip in which unusual weather spreads east across the tropical Pacific from Asia, is broadly possible months ahead. Many countries have mapped areas likely to be hit by floods and droughts. But they are often less clear about what to do with this information. In late 2001, leading climate agencies such as the United States government's National Oceanic and Atmospheric Administration (NOAA) had warned of a local impending El Niño. But on the face of it there was little planning going on. In Panama, for example, the local Red Cross said that there were no special emergency plans. "We will wait until it happens," said José Beliz in Panama. Similarly in Peru – where El Niño caused more than US$ 2 billion in damage to fisheries, crops and infrastructure in 1998 – the argument was that, while clearing up from a major quake was real, "right now El Niño is only a probability," said the Red Cross's Richard Medina. "We have no official warning and will only act when we do."

## Box 2.3 Risk mapping and relocation get political

Hazard risk mapping is an important first step in preparing societies for future natural disasters. The technique has been widely used in Latin America to identify zones at risk from floods, droughts and epidemics during El Niños. But, especially when accompanied by calls to relocate people from high-risk areas, mapping can also become an exceedingly political process. According to José Luis Rocha of the Central American University in Managua, the risk mapping that followed mudslides on Mount Casitas (which killed some 2,000 people during Hurricane Mitch) proved a money spinner for local landowners. Far from finding their land worthless, they successfully won compensation of US$ 3,000 per hectare – tens times its registered tax value – as aid agencies with plenty of money and not much time needed to relocate people.

The mapping of risk in southern Peru in the aftermath of the 2001 quake is proving a political minefield. The maps, being prepared with funding from the UN Development Programme, are "an important step to prepare for future disasters," says the provincial civil defence chief, Carlos Nacarino Rodriguez. Areas with loose soil, a high water table, suspect geology or on coastal land at risk from tsunamis have been identified as high-risk areas.

But Rodriguez admits there is widespread opposition whenever communities are asked to move, and suspicion that the authorities have ulterior motives. "Already there are complaints from the populations in these high-risk areas because the authorities won't let them rebuild," he says. And when communities oppose relocation in the aftermath of a disaster, aid agencies are left with a dilemma: help with rehabilitation and be accused of perpetuating risk, or refuse to help and stand accused of failing to prevent suffering? Two examples from southern Peru show what can happen.

The people of Catas, a small fishing and farming community where the River Tambo enters the Pacific Ocean, have been told to move. Half their village collapsed during the June 2001 quake. And geologists compiling hazard risk maps of the region say theirs is one of the most vulnerable to future quakes. Villagers are confused about exactly why. "They say the soil is cracked and sinking," says community leader Fernando Herrera. But in any case they suspect, rightly or wrongly, another motive behind why the authorities want them gone.

The village is a wreck today. The church collapsed. Of the 71 families in the village, 63 lost their homes. Three people died here. Most of the survivors live in tents and are fed from a charity food kitchen in the roofless remains of the village hall. In an unexpected twist, the quake raised the local water table so that it is less than a metre from the surface, and caused salty sea water to pour into the wells. So now they drink water from tanks trucked in.

Having categorized the village as a high-risk zone in a future quake, the municipal authorities have earmarked new land nearby for resettlement. But the villagers must pay the price for it, around US$ 40,000, themselves. If they don't, the authorities warn that the villagers will not get help with rehabilitation if they choose to stay. "For us this is a big dilemma," says the Red Cross's rehabilitation coordinator Freddy Gonzalez. "If they refuse to go to somewhere safer, should we help them? If we bring in housing modules, for instance, we would be encouraging rather than preventing a future disaster."

Herrera says they will agree to leave if the terms are right. They want to keep title to their old land. "We want to continue farming; we might build summer houses here by the sea," he says. "The trouble is the survey people say they found oil here. Some people think that is why they want us to go." Red Cross volunteers smile ruefully at the villagers' dreams and fear that, one way or another, they will be forced out just as soon as the oil derricks are ready to move in. "Risk mapping is a very political process here," says the International Federation's information delegate Fernando Nuño.

La Punta is a popular summer location for Lima's classier holidaymakers. The tsunami hit minutes after the June 2001 quake shook the continental shelf off southern Peru. The giant wave, some ten metres high, was triggered by the offshore tectonic shudder and swept towards the coast. It washed over the seafront and destroyed most of La Punta's buildings. It didn't stop until it hit the cliffs a kilometre inland.

Luckily it was winter. About 60 people died, mostly watchmen and their families, plus some farmers living close by on the coastal plain. "If it happened in summer, 8,000 to 10,000 would have died here," said Red Cross volunteer Carlos Franco, surveying the broken buildings and watching a handful of men who had returned to build the chalets, bars and brothels on which their income once depended. Watchman Raul Rojas stood beside the only thing left from his house, one block from the beach – a porcelain toilet bowl.

While Rojas has stayed behind in the ruins of La Punta, most of the permanent residents have gone. They now live on a deserted hillside eight kilometres from their old homes. Almost all of them are women and children living near the roadside in tents, most of them former farm labourers and maids of the holiday homes and hotels. This refugee encampment is called Alto Cerillo. "We are too scared to go back to our old homes, especially the children; we will never return," they said. Up here they have no permanent homes, only a weekly tanker to provide water. They lack access to schools or jobs or churches or markets, and are entirely reliant on food aid.

Are these people expected to stay here? "It's a temporary place till we find an appropriate permanent relocation," civil defence chief Rodriguez said. But city officials said the settlement was permanent. "The city is expanding. Eventually it will come out to meet them," said Alfredo Mezo of the charity Caritas, which is helping them.

Again the suspicion is that some of the land from which people have been removed "in their own interest" is being slated for new resort developments. Certainly the provincial mayor Enrique Gutierrez did not back up Rodriguez's claim that everybody should leave the resort zone because of its high-risk designation. The council won't help rebuild the homes of the poor who lived there, but nor will it stop the rich rebuilding. "We are not going to think that way," said Gutierrez.

Are the refugees really better off camped up on the hill rather than down on the shoreline? Mezo shrugs. "It doesn't really matter. Their homes are destroyed and their jobs are gone. This is their new life." ∎

Some believe this is not an unreasonable strategy, but José Aquino of CARE in Lima takes a different view. "Last time, we had a lot of problems getting materials, especially medicines to cope with epidemics, to the flooded zones because the roads were obstructed." The approach may need to be different depending on whether the likely impact is flood or drought. While floods happen suddenly, droughts have a cumulative effect more resembling a socio-economic crisis than a conventional natural disaster. Aquino's concerns may be justified – in late March 2002, NOAA's administrator warned that "the Pacific Ocean is heading toward an El Niño condition". Meanwhile Peruvian officials said anchovies normally present in Peru's cold coastal waters were being replaced by tropical species – a classic response to El Niño conditions.

## Quantifying risk

Risk mapping is an increasingly popular activity among civil defence planners. The idea is to identify places most in need of preparedness for disasters. And, in the worst places, to relocate communities. But what risks do you map, and how do you quantify risk? Three examples from Peru illustrate the point.

All around southern Peru, geologists have drawn up detailed maps of earthquake risks in the aftermath of the 2001 quake and the coastal tsunami that accompanied it. But for a lucky accident of timing, thousands could have died in the tsunami that struck the southern Peruvian resort of La Punta after the quake in June (see Box 2.3). The new risk maps now designate the resort a high-risk zone. But while quakes in the continental shelf off the Peruvian coast do occasionally cause tsunamis (the last hit this area of coast in 1873), these events are quite localized. There seems little basis for saying that this resort is any more at risk than any other low-lying stretch of the country's coastline. If La Punta is abandoned, maybe the whole coast should be.

But while tsunamis seem to attract undue attention, volcanic eruptions are largely ignored. The biggest city in southern Peru, Arequipa, sits in a valley surrounded by three large volcanoes. Locals say that one of them, El Misti, has been smoking regularly since the seismic convulsion that caused the quake. Could it be about to "blow"? It is 500 years since the last major eruption. According to vulcanologists at Indiana State University, it remains a "considerable hazard" to the city, which has extended up valleys that would carry any lava flow. A major eruption would not only rain debris and lava on Arequipa, it could also break a major hydroelectric dam on its slopes and unleash a tide of water on the city. Yet according to civil defence chief Rodriguez, El Misti does not feature in the risk mapping. "We don't have a plan for that. But maybe we will," he said.

Equally out of fashion is concern about the risk of floods and mudslides from glacial lakes high in the Andes. Back in 1970, in one of Peru's worst-ever disasters, a lake of melt water which formed at the foot of an Andean glacier burst its banks after a small earthquake and rushed down a mountain valley, engulfing an estimated 60,000 people, half of them in the town of Yungay.

After 1970, Peru's leading hydroelectric company, ElectroPeru, began to survey most of the country's glacial lakes. In 40 cases, it has employed engineers to siphon off the water from potentially dangerous lakes. But five years ago, says glaciologist Cesar Portocarrero, it stopped the work, arguing that it was a government responsibility. "This is really dangerous. With global warming rapidly melting the country's glaciers, the risks of a new disaster are rising. New lakes are forming all the time. We no longer have them mapped, so the risk of another big disaster grows all the time." One high-risk area, he believes, is among the glaciers around Salkantay mountain near the Inca ruins of Machu Picchu. Luckily no lakes burst there during the 2001 quake.

Furthermore, the mapping exercises that are carried out tend to focus solely on hazard risks, but are blind to the other socio-economic factors which influence the potential effects of disasters. Quantification of risks is not complete without assessing both the vulnerabilities and capacities of those populations exposed to natural hazards.

## Culture of risk reduction

Unfortunately, disaster mitigation and preparedness are in many ways still the "poor relations" of the aid world – neglected and under-resourced. But in a region where thousands lose their lives every year to recurrent disasters, where El Niño damage can regularly cost countries 10 per cent of their gross domestic product and where catastrophes such as Mitch put back economic development by 20 years, nations badly need more sophisticated coping strategies for disaster.

Reducing the deadly effects of disasters in Latin America has two aspects – one longer term, one shorter term:

- **Build risk reduction into every development plan and policy.** This long-term priority will reduce vulnerability to disasters great and small. Simply championing economic development and poverty reduction is not enough. Development may sometimes exacerbate disasters – by degrading the natural environment, for instance, or moving people from quake-proof shanties to quake-vulnerable high-rise apartments.
- **Invest more resources now into disaster preparedness.** Ensuring development policies are more risk resilient will take decades – but disasters will continue to hit the region every year. More investment in disaster preparedness initiatives is urgently needed in the short-term, to ensure all exposed communities are less vulnerable to disasters. Priority measures – often very inexpensive – to improve disaster preparedness include: risk and vulnerability mapping, disaster awareness

and education, early warning and evacuation systems, stockpiling relief materials, training in response skills, and planning from community to national to regional levels to ensure sound coordination of disaster response.

The example of Hurricane Michelle shows that protecting citizens from disaster has more to do with political will and good organization than with material wealth. "Cuba has lessons for the rest of us," argues Ben Wisner, because of its "investment in basic needs and social capital such as the training of neighbourhood activists [and] scientific capacity such as Havana's weather institute and public health services."

Disasters undermine social and economic development. To ignore the chance to invest more in disaster preparedness is to fail gravely those at risk, and will undermine their efforts to fight their way out of poverty.

A culture of risk reduction needs to cut across the activities of both the disaster and development professions, as well as vulnerable communities and their governments. Riches alone won't save anyone from disaster. Yet you can be poor and still be well informed and well prepared.

*Fred Pearce, the principal contributor to this chapter and Boxes 2.1 and 2.3, is based in London and writes on science, the environment and development for numerous publications. He is environment consultant for* New Scientist *magazine. Box 2.2 was contributed by Juan Carlos Villagran De Leon, scientific advisor for CEPREDENAC and designer of the Coyolate early warning system.*

## Sources and further information

Christoplos, Ian et al. "Re-framing risk: the changing context of disaster mitigation and preparedness" in *Disasters,* vol. 25(3), 2001, pp. 185-98.

Grunewald, F. et al. *NGO responses to Hurricane Mitch*. Humanitarian Practice Network Paper. London: Overseas Development Institute, 2000.

Lavall, Allan. "Prevention and mitigation of disasters in Central America", in Varley, Ann (ed.), *Disasters, Development and Environment*. New York: John Wiley, 1994.

Lindores, Douglas et al. *Review of major operations in the Americas: Hurricanes Georges and Mitch and Venezuela floods*. Geneva: International Federation, 2001.

Pearce, Fred. "Acts of God, acts of man?" in *New Scientist*, vol. 130, 18 May 1991, p. 20.

Rocha, Jose Luis and Christoplos, Ian. "Disaster mitigation and preparedness on the Nicaraguan post-Mitch agenda" in *Disasters*, vol. 25(3), 2001, pp. 240-250.

UN Office for the Coordination of Humanitarian Affairs (OCHA). *Preliminary Report: UN Interagency Mission in response to Hurricane Michelle passing through Cuba*, Geneva: OCHA, 2001.

Wisner, Ben. "Risk and the neo-liberal state" in *Disasters*, vol. 25(3), 2001, pp. 251-268.

## Web sites

Centre for the Prevention of Natural Disasters in Central America (CEPREDENAC) **http://www.cepredenac.org**

International Federation **http://www.ifrc.org**

NOAA **http://www.noaanews.noaa.gov/stories/s886.htm**

OCHA **http://www.reliefweb.int/ocha_ol/index.html**

Pan-American Health Organization **http://www.paho.org**

Reuter Foundation's AlertNet **http://www.alertnet.org**

UNICEF **http://www.unicef.org**

United Nations Intergovernmental Panel on Climate Change **http://www.ipcc.org**

chapter 3

# Preparedness pays off in Mozambique

The response to the 2000 floods in Mozambique – the worst for over a century – was in many ways a great success. For international observers, the disaster was epitomized by the iconic television image of the helicopter rescue of a mother who gave birth to a baby daughter while sheltering in a tree. Less reported were the 45,000 lives saved – the vast majority of them plucked out of the swirling waters by regional rather than international rescuers.

A year later, another wave of devastating floods hit a different part of Mozambique. Again, local teams, operating mainly by boat, rescued over 7,000 survivors. In each year, for every person who died, over 60 were saved. While media images of helicopters rescuing poor Africans gave the impression that international aid agencies saved the day, the real story is very different. Despite being one of the world's poorest countries, Mozambique was better prepared than many had feared. And, although international help was crucial, it succeeded because agencies let Mozambicans lead.

This chapter analyses both floods to identify which were the key factors in terms of disaster preparedness that ensured successful responses. What worked, what didn't and why? Following the 2000 floods, was reconstruction money invested in disaster preparedness? Were lessons learned during 2000 applied in response to the 2001 floods? What remains to be done to enhance disaster preparedness after the experience of two consecutive years of flooding? Recent experience in Mozambique would suggest that investing in disaster preparedness pays dividends in terms of lives saved. Yet the reality is that concrete progress in reducing the risks posed by disasters is being undermined by the cuts in public spending which Mozambique is being asked to make in order to integrate into the global economy.

## Record floods

In mid-January 2000, heavy rains fell earlier than usual, flooding rivers in southern Mozambique. The floods were not unusual, and were handled by local authorities. However, when, in mid-February, tropical cyclone Connie dumped record amounts of rain on the capital Maputo and the nation's southern watersheds, a full rescue effort began. Water in the Limpopo River was as high as in 1977 (the worst flood in living memory). The Mozambican navy, the Maputo fire brigade and the Malawian and South African air forces moved in to rescue stranded people. Hundreds of Red Cross volunteers took up their posts as tens of thousands of people fled the floods.

Photo opposite page:
Devastating floods hit Mozambique in 2000 and 2001. Local boat owners, together with the Mozambican military and the Mozambique Red Cross rescued 65 per cent of the 53,000 people saved.

Christopher Black/ International Federation, Mozambique 2001.

A few weeks later, another cyclone dropped more rain across the region. This third flood exceeded all expectations. At Xai-Xai on the Limpopo, water was three metres higher than any flood in the past 150 years. An area nearly the size of Belgium and the Netherlands combined was submerged. The Limpopo valley is flat, and people who fled to high ground assuming they would be safe soon found their small hills under water. A combination of helicopters from neighbouring nations and hundreds of small, local boats plucked people off rooftops and out of trees. Although 700 died, more than 45,000 people were rescued in the first three months of 2000. TV crews, already in Mozambique because of the February flood, filmed dramatic rescues – including that of baby Rosita Pedro, born in the tree where her mother sheltered.

In response to live pictures of rescues taking place, there was a huge outpouring of international aid – over US$ 100 million from individuals alone. At the peak of the response, there were 56 aircraft being flown by aid agencies and foreign air forces. In all, half a million people were forced from their homes and spread out over 200 locations, from small huddles on hilltops to resettlement centres accommodating entire towns. Relief efforts were hampered by another cyclone in mid-March. Even by May 2000 not everyone had made it home.

The following year, heavy rains deluged the region once again – this time hitting central Mozambique hard. The Zambeze River basin flooded from mid-February until mid-May, its waters peaking at 2.6 metres above flood level. Whereas the 2000 floods swelled rivers which had no major dams, in 2001 the two big dams on the Zambeze River coordinated their discharges so that the flood level stayed relatively stable. The government reported 113 killed in 2001's floods and 7,133 people rescued – most of those were saved by boats (see Figure 3.1). The disaster was less media-genic and, perhaps as a result, attracted less international aid and attention (see Box 3.1).

The relatively low death tolls and high numbers of people rescued during each flood suggest that disaster preparedness made a considerable difference to the quality of the response. So, which elements of preparedness worked well, which worked less well – and which steps could be taken to ensure more thorough preparedness in future?

## Early warning and evacuation

The naïve answer to flood threats is that people should live on higher ground. And after the 2000 floods, 43,400 families were resettled, according to the government. But flood plains provide fertile farmland and most farmers want to work on it and live near it. So people will have to continue living with floods (and droughts). That means prediction, early warning and evacuation systems are essential – as well as community awareness-raising to ensure that these systems work in practice.

Up to five months before each of the floods, African meteorologists predicted above average rains for southern Mozambique (in 2000) and for central areas (in 2001). Their work was based on computer models, plus analysis of La Niña and the sunspot cycle. Yet it remains an inexact art. Rainfall predictions cover a broad area, typically one-third of Mozambique, but floods and droughts are often in much smaller zones. Thus medium-term warnings, two or three months in advance, can be given, but significantly often, they will be wrong. In 1998, for example, the Mozambican meteorology office predicted a drought based on the anticipated effect of El Niño – yet rainfall was normal. So when, in September 1999, they warned people to prepare for a flood "like the one in 1977", not everyone believed them.

Short-term warnings, days or hours in advance, can be given for floods because there is some rainfall and river-level monitoring, supported by limited computer models. In March 2000, there was advance scientific warning of the magnitude of the third Limpopo flood crest. But Mozambican officials were extremely reluctant to warn the public of a flood worse than any seen by their parents or grandparents – even many officials did not believe it was possible.

While the art of prediction could be improved, there is a pressing need to ensure that all the links in the chain, from high-tech meteorology to low-tech warning and

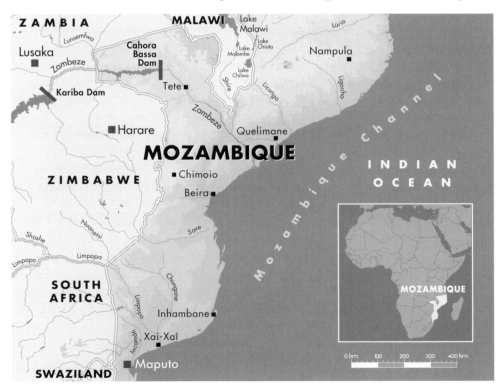

## Box 3.1 Media coverage – a double-edged sword

Spectacular images of heliborne rescues on live TV were beamed around the world during the floods of 2000. The subsequent influx of material and financial aid, culminating in pledges of US$ 470 million to reconstruct Mozambique, were at least partly due to this international media coverage.

During the 2001 floods, however, negative media coverage may have been partly responsible for a meaner international response. In 2001, floods hit the Zambeze valley – steeper-sided and less densely populated than the flat flood plain of the Limpopo which flooded in 2000. Along the Zambeze, many farmers live above – but farm fields below – the flood level. This led to the unexpected phenomenon of people declining to be rescued. Although their fields were under water, their houses were secure, they had brought their livestock near their homes and they had food stocks, so they chose to stay put.

The TV cameras came with the helicopters, but then went away when there were no dramatic rescues. On 17 March 2001, BBC correspondent David Shukman actually wrote in a leading European magazine that the unwillingness of people to be rescued meant it was "a bogus disaster" and Mozambique was "faking it".

Shukman also complained that "international teams had flown in to perform a high-profile rescue, but instead found themselves shuttling food around". However, the idea that international aircraft alone are responsible for disaster rescue is a media and aid-industry myth. Of the 53,000 people saved in 2000 and 2001, all but 4 per cent were rescued by Mozambicans and other African operators.

And 63 per cent of survivors were saved by boats.

While the large number of international planes that came after the flood were often too late to rescue anyone from trees or rooftops, they nevertheless saved thousands of lives, precisely because they transported food to those who were stranded. "Shuttling food" may not be high profile, it does not make good TV and it doesn't create such a warm glow in the hearts of donors – but it is what's needed in a flood like this.

However, the result in 2001 was less TV coverage, less donor interest, and therefore fewer planes and less food. By late March, with the press long gone, people were running out of food – and the river was not falling. More than 500 people a day, on foot and in boats, were making their way to accommodation centres. The roads in the Zambeze valley had turned from dirt to mud. Airlifts were the only way to get food to people, but there were just 20 aircraft to do the job. By May, there were 220,000 people in 65 centres. With less foreign aid, and especially fewer planes, conditions in the centres were not as good as during 2000. The ministry of health reported "severe nutritional problems" in some centres, and there were reports of cholera.

What can be learned from this experience of positive media coverage one year and negative coverage the next? Perhaps, as part of preparing for disaster, aid organizations could consider building closer relationships with journalists. Explaining to the media in advance the specific, unique contexts of each disaster could prove of positive benefit to both aid agencies and, more importantly, those at the receiving end of aid. ■

evacuation, are maintained. Those at risk from floods – who may often be the poorest and least educated – need to be convinced by people in whom they trust that it is time to leave, before the floods sweep them and their possessions away.

President Joaquim Chissano told an international conference in October 2000 that "warnings must be clear and simple". Improved warning systems, perhaps involving local leaders with radios or mobile telephones, were discussed. Chissano suggested using primary-school teachers as flood monitors – to watch rivers and issue warnings to their communities. Then people could receive clearer warnings one or two hours before the water reached their doors. This idea would also be cost-effective since it builds on community resources (e.g., teachers) which already exist and which are more likely to be trusted by local people (see Box 3.2).

Following two years of floods, the government began distributing radios, bicycles and motorcycles in areas at risk in late January 2002. The distribution started in Nampula province, where higher than normal rainfall was forecast. The equipment was given to members of local communities involved in disaster management to help monitor the situation and improve the flow of information. The government also distributed 16,000 posters with advice on how to prepare for a disaster.

However, even if people believe the warnings, what will they do with them? Impoverished Mozambicans must make careful choices about when to abandon their possessions. If you leave your house too soon (or unnecessarily) your goods could all be stolen and you may return to nothing. In some places during the 2000 floods, eye witnesses reported that even police looted abandoned homes. Fear of this happening meant many of the poorest people waited until the very last minute to flee, usually when they saw or heard the water. In flat areas like the Limpopo, that didn't leave them much time. Many families were afraid to abandon cattle and goats, and a significant number of the 700 who died were probably family members left behind to tend animals. Three options could be considered to tackle this problem:
- places of safety for cattle and goods;
- clearly marked escape routes and safety zones; and
- legal powers which force people to leave.

In most areas, there will be a week or more's warning of a serious flood. If the government or municipality could organize emergency cattle pens on high ground, and somewhere secure to store valuables like radios and cooking pots, then people could move their most precious property in advance and would be less reluctant to leave their homes. One of the disaster preparedness priorities for villagers in Inhambane province is the construction of a community strong-house where they can safely store their belongings during floods. But for much of Mozambique, this is

## Box 3.2 Do people heed warnings?

Manuel Jonasse is the head of the Mudamufe resettlement area in Sofala province. He arrived there with his wife and five children in February 2001 after the Púngòe River flooded. He said he'd received no warning of impending floods and was quite unprepared. "I've never seen floods like this year. It was the worst ever," Jonasse commented. "The secretary of the neighbourhood did go round warning people to move to higher ground, but the flood had already arrived by then."

Not all communities in the Púngòe valley were taken by surprise in the 2001 floods. Manuel Zalela used to live next to the Púngòe River in the Zona Ferroviaria. He said that, starting in December 2000, the head of the local administration called several community meetings "to warn that a big flood was coming". But the warnings seem to have had little effect. "Nobody believed it would happen," Zalela said. "I myself didn't believe it." After spending a night in the family's grain store, which is raised above ground level to protect it from pests, Zalela and his family were rescued by Red Cross boats and taken to safety.

Warnings were similarly ignored in the Buzi valley, also in Sofala province. Angelina Passe used to live at Chissamba on the banks of the Buzi River. "We heard the flood warnings, but we paid no attention because one of the old men said that floods had never arrived in that area. The flood came during the night. When we woke up, the house was surrounded by water, but it was not flowing very fast." Angelina and her neighbours fled. "There were about 20 of us perched on the nearest high ground – we stayed there all day. In the evening a boat from the local administration came to rescue us."

When floods are a regular event, an early warning system is indispensable – and its reliability is essential if people are to heed the warnings. A possible solution in the context of Mozambique is to use teachers as flood monitors. This has several advantages. There are many teachers and they are respected in their communities. They are underpaid and would welcome extra income for doing community disaster education and for being "on call" during the flood season. The teacher-monitors could be part of a river-watch system in which they regularly check the height of the rivers with gauges, issue local warnings and report their findings to a district coordination centre. ■

simply too expensive to contemplate – in part because high ground is often 20 or 30 kilometres away and beyond the reach of community initiatives.

Yet poverty is not in itself a barrier to better preparedness. In Bangladesh, for example, where vast floods occur almost annually, the Bangladesh Red Crescent has helped villagers build raised earth mounds where they and their animals can take refuge during floods. So far, nothing like this has been considered in Mozambique. Such projects require a level of collective community action which, while evident in rescuing flood victims, subsides in the downtime between disasters.

One cheap way to spread disaster awareness through communities, and therefore improve their chances of survival, is to clearly signpost previous flood peaks, escape routes and safe refuges. Painting on lamp posts and walls the height of flood waters in, for example, 1977 or 2000 would allow future warnings to be keyed to memories of past floods. As Ian Davis, visiting professor to the UK's Cranfield University's disaster management centre, has pointed out, schoolchildren could do this during their geography lessons, and the cost would be "as much as a pot of paint".

Escape routes and safe refuges need to be identified by local people, based on their own experience and local knowledge. The government agency responsible for disaster management is canvassing for funds to build refuge platforms in the Limpopo and Save River valleys, where virtually entire resident populations were displaced by flood waters in 2000.

Legal powers also play a key role in improving disaster preparedness, and Mozambique's government has approved a bill on disaster management to be presented to parliament. If passed, this law will allow the president to declare a state of emergency in the event of disaster and will enable the cabinet to order the compulsory evacuation of areas at risk.

## Practice improves preparedness

When, in late 1999, meteorologists warned of unusually heavy rains likely to hit southern Mozambique the following year, the Mozambique Red Cross (CVM – Cruz Vermelha de Moçambique) immediately began to retrain volunteers in the areas likely to be affected. The government's disaster management institute sent teams out to prepare and to warn people in vulnerable areas that floods could be as bad as record ones in 1977. The ministry of health (which normally delivers medicines quarterly to provincial clinics) dispatched January's shipments in December, to ensure adequate supplies in case of floods. Meanwhile, senior officials were sent out to confirm that health posts had adequate stocks to treat malaria and cholera. Government officials even cancelled their usual December/January holidays to be ready for the floods. And they renewed contacts with the South African air force, which had helped in previous emergencies.

All in all, Mozambique's own disaster management agencies were reasonably well prepared. But, some observers have asked, how could this happen in such a poor country? The answer lies in Mozambique's past experience of disasters and conflict. During the colonial era, the Portuguese authorities paid little attention to floods and droughts, because they mainly affected the "indigenous" population. At independence in 1975, the new government quickly had to respond to three years of serious floods and a major El Niño drought in the early 1980s. The young government set up a department to prevent and combat natural disasters (DPCCN) in 1980; the CVM was established the following year.

Through the 1980s, Mozambique became a cold war battlefield, and western European countries poured in huge amounts of relief. As a result, says the CVM's secretary general Fernanda Texeira, this "accumulated experience from the emergency years, both in the Red Cross and within communities, contributed to the success of operations in 2000 and 2001".

The war ended in 1992, followed by elections in 1994. Both the DPCCN and the Red Cross needed to rethink their roles and in 1999 they restructured to concentrate on natural disasters, while drawing on those members who had gained emergency experience during the decade of war. The DPCCN was resurrected as the national disaster management institute (INGC), and decided to concentrate on coordinating other agencies providing disaster relief. The Mozambique Red Cross, meanwhile, opted to concentrate on providing basic health care for isolated and displaced people in emergencies, and began training volunteers.

Limited floods in early 1999 gave both organizations experience in their new roles. In late 1999, the INGC ran a major simulation exercise in flood relief, involving the police, CVM, local flying clubs, fire brigade and scouts (continuing a practice established nationwide since 1995 of simulating response to different types of disasters each year). Thus both the CVM and INGC went into the 1999-2000 rainy season with some degree of readiness.

## Stockpiled supplies too tempting

One of the key features of the restructuring of the DPCCN and the CVM was the recognition that stockpiles of relief supplies, while attractive in theory, provided too much temptation in practice. Mozambique has bitter experience of emergency reserves of food, fuel and vehicles being stolen or sold by corrupt officials, and salaries are so low that this seems unavoidable. Speaking in Maputo at a seminar on lessons learned from the floods in 2000, Mozambican foreign minister Leonardo Simão commented that during the 1980s, "thefts and pilfering were constant. Some staff thought they had the right to steal. DPCCN could not be reformed. So we had to create a new institution with a new conception. There would be no lorries or warehouses, because that is the basis of theft and misuse."

In addition, the gap between major droughts or floods is sometimes a decade or more – maintaining stocks in good condition for that long a period is almost impossible under Mozambique's climatic and economic conditions. As a result, both organizations slimmed down their large warehouses and transport fleets.

Yet one of the most important lessons learned from both floods is the need for stocks that can be drawn on quickly in case of emergency. Prior to both floods, the United Nations' (UN) World Food Programme (WFP) and children's agency (UNICEF) pre-positioned food and medicines, based on disaster predictions. And the CVM, along with other international Red Cross partners, also had relief stocks left over from 2000 in place for the 2001 floods. So while stockpiling supplies over long periods may not work, nevertheless agencies can ensure that relief stocks are requisitioned in time to ensure they can be moved to vulnerable areas before disaster strikes.

Francisco Pateguana, former governor of Inhambane province (hit hard by 2000's floods) argues for relief contracts or retainers, on the grounds that private companies are more likely to keep supplies in good condition if they continue using them for their own purposes afterwards. For example, the local petrol station would be paid a fee to maintain a stock of fuel during the annual three-month flood season. If there was a flood, the government would buy the fuel. But if there was no flood, the fuel would be sold off in the normal way over subsequent months. Similarly, boat owners would be paid a retainer if they agreed to supply a boat and pilot at 24 hours notice, if requested during the flood season.

More than 200 inflatable boats, donated by international agencies, were used to rescue people and transport food in both 2000 and 2001. Indeed, boat rescues during

Even when flood waters reach the village well, people are reluctant to abandon their homes, livestock and belongings.

Christopher Black/
International Federation,
Mozambique 2001.

2001's floods accounted for 4,400 of the 7,133 lives saved (see Figure 3.1). There was a debate about whether to reserve the boats for emergency use only. Arguments in favour are obvious. However, if kept in storage for long periods in tropical conditions, rubber boats would deteriorate and be attacked by rats. Also, since Mozambique is so poor and so short of river transport, using the boats as ferries would stimulate the economy. In practice, the boats were used commercially in between disasters to cross rivers where bridges or roads had been washed away.

## Coordination works when locals lead

One of the most striking lessons from both years of flooding is that coordination of the relief effort worked best when Mozambicans led – or fully participated in – all aspects of the disaster response. In the first days of 2000's floods, before outside help arrived, internal organization was key. Local health workers and the CVM set up emergency health posts. Local government officials organized the resettlement centres; in Chiaquelane centre, for example, people were told to congregate according to their home neighbourhoods, and local leaders took charge of distributing tents and food, and constructing latrines and water tanks.

As 2000's floods became more severe, foreign minister Leonardo Simão personally took over coordination of the relief effort. The UN's office for the coordination of humanitarian assistance (OCHA) sent a team which, rather than being located in a UN office, was set up in the INGC's offices, in order to enhance the joint disaster coordination centre.

Thousands of people from 250 different organizations came to help. UN agencies alone had 500 people working on flood relief. Seven National Red Cross Societies sent teams. As aid flooded in, coordination was essential to prevent chaos. At both national and local levels there were daily meetings, usually chaired by the government. Individual agencies were given responsibility for tasks, such as food distribution or water supply, in a particular camp or area.

With large volumes of people, aircraft and aid to coordinate, a joint logistics operations centre (JLOC) was established at the INGC office in Maputo, linked to a logistics cell at the airport. Requests for transport of food, medicines, tents and staff were submitted daily to the JLOC, approved and prioritized, and sent to the airport where an afternoon meeting assigned tasks to available aircraft. For the first time during a natural disaster in Mozambique, nine military air forces accepted tasking by a civilian coordination system, rather than deciding individually what to fly and where.

The effect of the coordinated effort was spectacular. Adequate water and sanitation meant there was no cholera; sufficient health staff kept malaria under control; enough food meant there was little hunger. Overall, the death rate of displaced people was lower than if they had stayed at home.

Yet strangely the great success of the joint logistics operations centre in 2000 was not repeated in 2001. During the latter flood, the JLOC was based up the coast in Beira and fewer agencies participated. By 2001, Mozambique had two military helicopters in service. But neither they, nor the South African air force, nor the aircraft hired by INGC worked through the JLOC. Allegedly management problems – on the part of both the INGC and the UN – were to blame. According to one UN official, the magnitude of 2000's floods meant that contingency planning for 2001 started late and this in turn contributed to the weakness of the Beira-based JLOC. According to a Red Cross official, the government's provincial staff in Beira were far less competent than those based in Maputo, which had

Figure 3.1
Africans saved
Africans

Source: Christie &
Hanlon (2001); INGC

| Operator | 2000 | | 2001 | |
| --- | --- | --- | --- | --- |
| | Air | Boat | Air | Boat |
| Mozambican military | | 17,612 | 1,850 | 3,100 |
| South African military | 14,391 | | 357 | |
| Mozambique Red Cross | | 4,483 | | |
| Malawian military | 1,873 | | | |
| Air Serv (international NGO) | 208 | | | |
| French military | 79 | | | |
| Portuguese Civil Protection | | | | 1,300 |
| Fire service, private boats | | 7,000+ | | |
| Miscellaneous rescuers | | | 526 | |
| **Total rescued, at least** | **16,551** | **29,095** | **2,733** | **4,400** |
| **Total by year** | **45,646** | | **7,133** | |

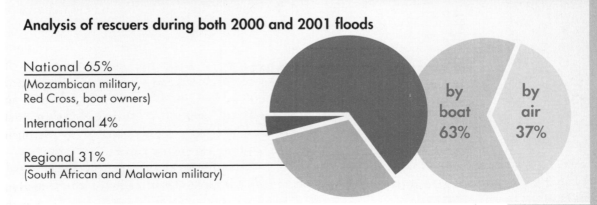

## Analysis of rescuers during both 2000 and 2001 floods

National 65%
(Mozambican military,
Red Cross, boat owners)

International 4%

Regional 31%
(South African and Malawian military)

by boat 63%

by air 37%

formed the hub of 2000's relief operation. As a result, with fewer planes in the pool and poorer coordination, the airborne element of 2001's disaster response was less effective than in 2000.

One lesson from two years of flood relief operations is that coordination mechanisms remain very fragile, based more on personalities than structures. A key aspect of improved disaster preparedness in Mozambique would be to enable the INGC to coordinate disaster relief throughout the country, by providing it with well-trained staff and resources, as well as to require all aid operators to work through it.

The Red Cross system proved particularly effective during both years of flooding – and this too was the result of Mozambicans preparing for and fully participating in disaster response. The Mozambique Red Cross worked directly with Red Cross societies from other countries, which took responsibility for various aspects of relief and reconstruction. During 2000's floods, 683 CVM volunteers assisted over half of those displaced. CVM's secretary general Fernanda Teixeria stressed that material aid from overseas Red Cross societies matched local needs: "They know what is needed in a disaster. They sent what we asked for – tents, soap, kitchen sets." And they sent money quickly, so that CVM could buy locally what it needed urgently.

During 2001's floods, provincial Mozambique Red Cross coordinators played an even bigger role and were given more power than in 2000. The International Federation of Red Cross and Red Crescent Societies, while maintaining a significant presence in Mozambique during 2001, worked in CVM offices and often through CVM's provincial coordinators. Mark Wilson, the International Federation's head of delegation in Mozambique at the time, points to improvements in the Red Cross response in 2001: "We had stockpiled 4,000 family relief kits. We had trained more volunteers through disaster preparedness programmes, and crucially, working with the CVM, we were able to manage and coordinate international Red Cross responses so that the CVM was not submerged. We were able to gradually expand our response to meet growing needs in a seamless way."

## Continuity key to success

Not surprisingly, international agencies found that their ability to respond to 2001's disaster was greatly improved if their staff had had prior experience of dealing with disasters in Mozambique. UNICEF's provision of water and sanitation for people in temporary centres during 2001 benefited from experience gained the previous year. Jonathan Caldwell, UNICEF emergency coordinator during both floods, observed that, "Operations were much better in 2001 because of existing staff with experience. This was one of the lessons taken over from 2000 to 2001."

## Box 3.3 Community-based disaster preparedness

Matasse is a rural community of some 2,000 people threatened by flooding from the nearby Save River. Last year, the Mozambique Red Cross (CVM) undertook a pilot project there in community-based disaster preparedness (CBDP) with support from the Danish Red Cross.

CVM emphasizes the importance of respecting local tradition and involving community members in data collection, risk mapping and planning, if such projects are to succeed. So, after making contact with the district authorities and local Red Cross committee, the team approached Matasse's headmen to explain the purpose of the proposed project. Community meetings followed to describe the project and recruit volunteers.

The community volunteers were then trained to analyse potential hazards and identify ways of preparing the community to save lives and secure livelihoods. Volunteer training also included first aid and methods of community education in HIV/AIDS and land-mine awareness.

This training was soon put to use in drawing up a history of disasters and a seasonal calendar of the area. The earliest disaster in the collective memory was flooding in 1939, followed by a pattern of drought and flood up to the present, including floods in 1999 and 2000. The disaster history also recorded how people coped in past disasters. The seasonal calendar indicated periods in the year when the population was most vulnerable to poverty and health risks.

The Red Cross team then led a transect walk with community members – literally walking in a straight line across the area at risk, visually identifying its physical features. On the basis of this, they mapped existing resources, infrastructure and possible risks and hazards, plotting details with a GPS (satellite-based global positioning system). Risk maps were then created using GIS (geographic information system) technology introduced by the Danish team members. The maps covered residential and farming areas and identified those most at risk from flooding, as well as the best places of refuge.

Community involvement in the project helped to identify a series of objectives relevant to the real situation in Matasse and the priorities of its population: planning of mitigation activities; recruitment and training of new volunteers; improvement of wells; participation in rescue training; and distribution of radios to improve early warning.

Priority mitigation activities include planting trees to halt erosion near the riverbank, and constructing a multi-purpose community hall in a secure location to serve as a store for pre-positioned relief stocks and household goods in the event of disaster. The hall would also serve as a community meeting centre.

Conscious that one organization or community alone cannot bear the burden of disaster preparedness, CVM is mobilizing support from other agencies for these activities. As a result, CARE is cooperating in the improvement of wells, the International Federation is helping with rescue training, and FEWSNET (Famine Early Warning System Network) is providing radios to Red Cross volunteers. ∎

Conversely, a rapid changeover of staff risks reinventing the wheel. One failure during the 2001 floods was the inability to set up data collection and management mechanisms that would allow coordinators to make informed decisions on the allocation of resources. Non-governmental organizations (NGOs) and UN agencies were critical of the emergency team hired by the UN secretariat in Mozambique to staff its emergency coordination unit, which did not have enough knowledge of the 2000 operations. UNICEF's Caldwell said: "The database failed because too much information was being collected and it was too late. And the form for collecting information in assessments was changed without consultation." Some observers argue that the informal exchange of information during regular coordination meetings in the 2000 floods worked better than the more complicated data management system attempted by the UN in 2001.

Improved disaster preparedness requires drawing on a larger pool of people. Both floods showed how quickly local and provincial Mozambican officials moved to participate in rescue and relief, especially in organizing accommodation centres. Much could be done to improve their training. And more could be done through local mobilization. For example, while the INGC may be strong at national level, it lacks effective representation at provincial and local levels. Although it may not be practical to employ full-time staff at lower levels, existing civil servants could be given emergency responsibilities, additional training and even small pay top-ups when, for example, the president declares a state of emergency.

At the community level, the Mozambique Red Cross showed that people are prepared to volunteer and is committed to training volunteers for future disasters. "It is an investment that pays off," says CVM's Texeira. "During the 2001 floods in the Zambezi valley, I went to visit Sena district and found there a group of volunteers trained in the drought of 1992-93 who were working in the relief effort." Their training is simple and includes instruction in how to erect tents, organize a camp, register displaced people, assess needs, chlorinate water, build latrines, as well as carrying out first aid and boat rescues. The advantage of such broad community-based disaster preparedness training is that it can be applied to a range of different disasters (see Box 3.3).

## Mitigation measures

There is not scope within this chapter to examine the full range of structural and non-structural means of disaster mitigation in Mozambique – for example, an analysis of the effects of land use and river basin management on flooding. However, several mitigation measures currently being implemented in the region are worth mentioning.

During 2000, road embankments were seen to trap water and extend the duration of floods in the Limpopo valley. As a result, more gaps and bridges are now being built into the embankments to allow flood waters to pass underneath.

## Box 3.4 New homes and livelihoods reduce disaster risk

Luisa Fabião Macie has three grandchildren, but she's not sure of her age. One day, the secretary of her neighbourhood in the Incomati valley went round warning people that floods were on the way. Luisa was in the middle of packing up her belongings when the swirling flood took her by surprise and her packed belongings with it.

With her two children, Luisa spent two days in a tree. On the third day, a helicopter came to their rescue. "I'm not sure how I went inside the helicopter," said Luisa. "It all happened so fast, then suddenly I was inside and so were my children."

The floods which swept through the valley left thousands like Luisa homeless – often with only the clothes they wore when the torrent forced them to flee. They spent long months in temporary accommodation centres. Slowly, resettlement projects began to rehouse them. Every household made homeless is, in principle, eligible for a new home safe from the danger of floods. NGOs have helped establish these new rural residential neighbourhoods. So far, with floods still fresh in the memory, local officials have found no resistance to relocation.

Luisa now lives in Taninga, a resettlement area established last year above the Incomati valley, three kilometres from her old home. When the British NGO ActionAid began its resettlement project, Luisa qualified for a new house, one of hundreds now lined up in neat rows. Each house features cane walls, a wooden door, a corrugated zinc roof, and two windows complete with mosquito nets and shutters.

A new house was not the only change. Luisa now works in a project funded by ActionAid to introduce a new variety of sweet potato – one with orange flesh, indicating a high content of vitamin A. The NGO pays a group of women to cultivate the plants, and when enough are ready, they are distributed to small farmers in the surrounding area. Luisa has also branched out into trade. She uses the money earned from growing sweet potato to carry out small-scale trading in household items, such as sugar and salt.

This kind of resettlement, as well as ensuring families recover quickly from their ordeal, reduces the risks posed by future disasters. The new homes are not only more flood-resistant, but their mosquito nets protect inhabitants from malaria. And, crucially, investing in the livelihoods of the new community ensures that they remain in a safer area, rather than returning to the dangers of their old homes and land. Reducing the risks posed by flooding means not just building stronger houses, but also building stronger livelihoods, so that the most vulnerable can start to save money and protect themselves from future shocks. ■

Investment in flood-proofing vital public infrastructure such as health posts, schools and government offices pays off. Well-constructed clinics survived the floods with little damage, suggesting that present design standards for some government buildings are already adequate. Making a flood-proof community strong-house, in which to store valuable possessions, would also encourage more people to evacuate before flood waters carry them off.

Some NGOs are providing building materials to enable those who lost their homes to rebuild in a more flood-resistant way. In resettled communities such as Taninga, this structural mitigation is combined with non-structural initiatives such as reinvigorating small-scale trading and farming (see Box 3.4). Well-planned recovery can thereby create a virtuous spiral, which reduces the risks posed by future disasters.

Another non-structural mitigation measure, practised in neighbouring South Africa, is to create artificial small floods every few years by controlled releases of water from dams upstream. These minor floods serve two purposes: first, to flush out from river beds, banks and basins the silt, undergrowth and dead trees which otherwise obstruct the flow of a major flood and make it worse. Secondly, such controlled floods would encourage people to move their homes further away from the edges of river banks and create a greater, more continuous awareness among vulnerable communities of the dangers of flooding.

## Rhetoric/reality gap

The first step to improved disaster preparedness is better data, particularly on rainfall and river levels, and improved computer models to analyse that data. This would allow earlier and more accurate warnings. There is substantial donor rhetoric about improving warnings, but donors proved exceedingly reluctant to come up with the money. Of the money the Mozambique government requested simply to replace river and rain gauges destroyed by 2000's floods, donors promised just 15 per cent. It took more than a year to negotiate money for their replacement. This was despite the fact that in May 2000, donors pledged US$ 470 million in recovery and reconstruction aid. Other key works, such as repairing dykes before the next rainy season, were not possible because of the slow release of donor funds.

Once the disaster was over, donor agencies again pursued their own, rather than Mozambique's, priorities. So more money than Mozambique requested was offered in support of private farmers and for water and sanitation, while flood warnings and repairs of public buildings like town halls were not funded because they did not meet donor priorities. In some cases, it took months of negotiation to convince a donor to move away from priorities set at headquarters level, even when their own office in Mozambique admitted the need to change. While donors mouthed the rhetoric of "local empowerment" and "risk reduction", in reality Mozambique's ability to set disaster reduction priorities lost out.

Where donors do invest in disaster preparedness, the lack of coordination with the Mozambique government, or even with each other, has led to too many training seminars for already overworked civil servants. And with many civil servants forced to take extra jobs in order to earn enough money to feed their families, it is not clear how much more can be demanded of them. So, although Mozambique has a serious skills shortage, the answer is not simply for aid agencies to run ever more training sessions which leave officials too little time to do their actual jobs.

Following 2001's floods, risk reduction did climb up the agenda somewhat. Finland, Germany, Ireland, Japan, the Netherlands, Russia and the United States supported disaster preparedness activities through the INGC and other institutions. However, the main donor to the INGC is Italy, whose pledge at the 2000 Rome conference was still tied up by bureaucracy in February 2002. "We need to see greater donor flexibility in using rehabilitation money to improve preparedness," argues Eva von Oelreich, head of the disaster preparedness department of the International Federation.

## Risk reduction undermined

Ultimately, the obstacle to most forms of disaster preparedness and mitigation in Mozambique is the acute lack of financial resources. Even keeping the early warning chain operational needs money to pay and train flood warning monitors, and to provide key coordinators with basic resources like bicycles, radio batteries and, possibly, mobile phones. Not large amounts of money, but it would mean an increase in government spending at a time when structural adjustment tightly caps the budget.

Mozambique's poverty reduction strategy paper (PARPA – Plano de Acção para a Redução da Pobreza Absoluta 2001-2005), led by the World Bank and the International Monetary Fund (IMF), keeps tight control on government spending. It calls for a cut, in real terms, in spending on "priority areas" for poverty reduction, from 19.4 per cent of gross domestic product in 2001 to 17 per cent in 2005. Faced with having to slash social spending, the Mozambique government has made the difficult decision to invest more in health while making cuts in education over the short term. In 2002, education spending is being cut by an incredible 12 per cent. No provision, for example, is made to train replacements for the nearly 900 teachers a year who may die of AIDS.

Under such severe spending limits, investment in comprehensive disaster preparedness and mitigation has lost out. The PARPA has a section on "Reducing vulnerability to natural disaster" but this simply says that natural disasters must be treated as a "risk factor" in forecasts of economic growth. It calls for the promotion of "a contingency plan for natural disasters" and for strengthening the capacity of the national meteorological institute. But no money is allocated for this. The plan calls for the establishment of a "flood warning management office", to be included in the category of "management of water resources". Yet spending in this category is to be cut by a staggering 90 per cent to just one-tenth of its 2001 level, with the money instead diverted to urban water supply and sanitation.

Although the PARPA contains all the right words about reducing vulnerability, the limits on state spending which it imposes have effectively forced Mozambique to choose between investing either in disaster reduction initiatives such as flood management, or in more developmental initiatives, such as better health care nationwide, and water and sanitation in urban areas. Not having the money to invest in all these initiatives, Mozambique opted for more immediate gains for the sick and the urban poor. Yet, this state of affairs contradicts the World Bank's own purported aims, as Swedish development analyst Ian Christoplos points out: "The PARPA reduces the government's scope for spending on disaster risk reduction, despite the fact that the World Bank is preaching internationally that 'security' – including security from natural hazards – is one of the three pillars of poverty alleviation."

## Africans rescued Africans

Several compelling lessons about disaster preparedness emerge from Mozambique's two years of record floods:

- **Early warning needs trust.** Predicting the weather correctly is only half the battle. Many Mozambicans did not believe the warnings. To maintain the early warning chain from high-tech meteorology to low-tech reaction means involving trusted community members, such as teachers, in the warning process. Red Cross experience shows that disaster awareness and trust can be improved by involving the community in data collection, risk mapping and disaster planning.
- **Evacuate quicker.** Many left it too late to flee the floods, for fear of their possessions being looted or their livestock being killed. More lives could be saved by investing in flood-proof strong-houses and cattle pens, where possessions could be stored, and animals moved, in advance. Marking previous flood levels, evacuation routes and safe havens would help increase public disaster awareness and save lives.
- **Agency preparedness pays off.** Following predictions of heavy rain, Mozambique's relief agencies organized a major flood simulation exercise before disaster struck – ensuring they had experience in working together. They retrained volunteers and, with UN agencies, pre-positioned essential relief supplies. They concluded that contracting local companies to provide emergency supplies would discourage the corruption associated with stockpiling aid in warehouses. However, closer collaboration and planning are needed between the government and its emergency services, Red Cross, UN and NGOs to resolve procedures well before disaster strikes.
- **Coordination works when Mozambicans lead.** During 2000's floods, Mozambique's foreign minister led the response. And a joint UN/government

logistics centre situated within his disaster management office ensured that all civilian and military relief assets – whether foreign or domestic – were efficiently tasked. In 2001, government coordination was managed at a provincial level and proved less effective than in 2000. Building the government's capacity to coordinate disaster relief assets – at a provincial as well as central level – is a key disaster preparedness initiative.

- **Africans rescued Africans.** During the rescue phases of both floods, around 53,000 Mozambicans were saved from drowning. Two-thirds of all those saved – 34,000 people – were rescued not by international teams but by Mozambique's own military and Red Cross. Counting the support of air forces from South Africa and Malawi, 96 per cent of all those saved were saved by regional assets. While international aid was crucial in supplying relief aid to survivors in resettlement centres, it arrived too late to play a major role in the rescue phase.

- **Training volunteers works.** None of the above disaster preparedness measures have a chance of working unless the right people are trained for the job. During Mozambique's floods, Red Cross volunteers, trained in disaster response as long as a decade earlier, still remembered their skills and put them to use. Investment in training local people will pay off – they will be there for the next disaster, while many international relief staff will not.

- **Gap between donor rhetoric and reality.** Donors mouth the language of risk reduction, but their words are not matched by the money needed to make it a reality. The IMF- and World Bank-led poverty reduction strategy, while promoting the idea of flood management, has, by limiting state spending, effectively forced Mozambique to choose between rural or urban risk reduction. Yet, until donors prioritize nationwide risk reduction as a key component of poverty reduction, disasters will continue to inflict loss of life, shatter livelihoods and undermine development.

Mozambique's response to two years of record floods was better than any outside agency believed possible. Many lessons were learned. However, the country's extreme poverty, combined with the unwillingness of donors to match risk reduction rhetoric with money and policy changes, make it very difficult for Mozambique to convert hard-won experience into more effective disaster preparedness.

*Frances Christie and Joseph Hanlon, journalists and authors of* Mozambique and the Great Flood of 2000 *(James Currey, Oxford, 2001) are the principal contributors to this chapter.*

## Sources and further information

Christie, Frances and Hanlon, Joseph. *Mozambique and the Great Flood of 2000.* Oxford: James Currey, 2001.

Cruz Vermelha de Moçambique (CVM). *Relatório Cheias 2000.* Maputo: CVM, 2000.

CVM. *Relatório Cheias 2001.* Maputo: CVM. 2001.

Government of Mozambique (GoM). *Mozambique Post-Floods Reconstruction Programme.* Maputo: GoM, 2000.

Government of Mozambique. *Emergency 2001 International Appeal by the Government of Mozambique in Collaboration with the United Nations.* Maputo: GoM, 2001.

Government of Mozambique. *Balanço do Apelo de Emergência face às cheias 2001.* Maputo: GoM, July 2001.

Government of Mozambique. *Post-flood Reconstruction Programme for the central region of Mozambique.* Maputo: GoM, July 2001.

Hanlon, Joseph. *Peace Without Profit: How the IMF Blocks Rebuilding in Mozambique.* Oxford: James Currey, 1996.

INGC. *Estruturas e estratégias de gestão de calamidades: política nacional de gestão de calamidades.* Maputo: INGC, 1999

International Federation. *Annual Report: Mozambique.* Geneva: International Federation, 2001. Available at http://www.ifrc.org

Office of the UN Resident Coordinator. *Mozambique Floods, final report.* Maputo: UN, 2001.

Simkin, Peter. *2000 Floods in Mozambique — Lessons Learned from the Humanitarian Relief Operation.* Maputo: UN System in Mozambique, 2000.

Torcato, Maria de Lourdes. "Chuvas e Cheias", in *MoçAmbiente,* April/May 2000, Maputo.

UN Secretary-General. *Assistance to Mozambique Following the Devastating Floods.* A/55/123-E/2000/89. New York: 2000.

UN Office for the Coordination of Humanitarian Affairs (OCHA). *Final Report, Workshop on Lessons Learned from the 2001 floods.* Maputo: UN Development Programme, July 2001.

## Web sites

Cruz Vermelha de Moçambique (Mozambique Red Cross)
   **http://www.geocities.com/TheTropics/6020/cvm_02.html**
International Federation of Red Cross and Red Crescent Societies report on
   Mozambique: **http://www.ifrc.org/where/country/cn6.asp?countryid=120**
Moçambique on-line Cheias em Moçambique:
   **http://www.mol.co.mz/cheias/index.html**
Mozambique National Institute for Disaster Management and Coordination:
   **http://www.teledata.mz/ingc/**
United Nations System in Mozambique reports on the 2000 floods:
   **http://www.unsystemmoz.org/news/flood/defflood2000.asp**
United Nations System in Mozambique reports on the 2001 floods:
   **http://www.unsystemmoz.org/news/flood/default.asp**

Section One

**Focus on
reducing
risk**

# Pacific islands foretell future of climate change

When the people of Tuvalu, an island nation in the Pacific, first encountered Europeans in the 19th century, they gave them the name *palangi*. Victorian travellers translated the word to mean "heaven bursters". The name has an uneasy resonance, now that climate change, a bad spell escaped from the Pandora's box of the rich world, is disrupting the lives of the region's people.

Stretched over 600 kilometres of ocean, Tuvalu is a string of nine coral atolls no more than a few metres above sea level at their highest point. Since their ancestors arrived around 2,000 years ago, the islanders have adapted to cope with living in one of the earth's most fragile and extreme environments. But Tuvalu's way of life, along with that of other low-lying islands and coastlines around the world, is now threatened by both a changing global climate and misguided development priorities.

During spring tides, the airport runway floods on Funafuti, Tuvalu's main administrative centre. Last year, for the first time in living memory, it flooded for five consecutive months at high tide. As a result, this tiny Pacific nation of 10,000 people attracted media headlines worldwide for the dangers it faces from rising seas. The government has asked neighbouring countries to help relocate its people. In March 2002, the prime minister announced he was considering taking the world's worst polluters to court (see Box 4.1).

But while scientists debate the precise nature of sea-level rise around Tuvalu, there are many other equally serious risks posed to all Pacific islands by global warming. Climate change threatens to make cyclones more intense and unpredictable. Storm surges associated with cyclones are already eroding coastlines and contaminating freshwater supplies. Rising temperatures are increasing the spread of infectious disease vectors. Droughts in some regions are predicted to become more frequent. The latest data show that the number of people in the Oceania region affected by weather-related disasters has soared by a staggering 65 times over the past 30 years (see Figure 4.1).

These disasters threaten to make life unviable for many coastal communities long before they finally succumb to rising tides. Still rusting in Funafuti's lagoon is the wreck of a large fishing boat that fruitlessly sought shelter during a cyclone in 1972. The island was flattened. Miraculously only a handful of people died. To avoid being blown out to sea many tied themselves to trees. The island recovered. But during the 1990s Tuvalu suffered seven cyclones. The United Nations (UN) Fiji-based regional disaster expert, Charlie Higgins, argues that, because of climate change, "the frequency

Photo opposite page: From 1992-2001, Kiribati's population was the fourth most disaster-affected on earth. Low-lying islands and coastlines are especially vulnerable to the weather-related disasters associated with global warming.

Caroline Penn/Panos, Kiribati.

## Box 4.1 Relocation – the last resort

On 5 March 2002, the prime minister of Tuvalu, Koloa Talake, announced at the Commonwealth Heads of Government Meeting in Australia that his nation was planning to sue the world's worst greenhouse gas polluters at the International Court of Justice, as predicted by the *World Disasters Report 2001*. According to the *Pacific Islands Report*, he stressed that global warming was an issue threatening both his people and country. "It is frightening," said Talake, "islands that used to be our playgrounds have disappeared. Some scientists say there is no rise in sea level, but the tide is rising. We have seen it with our own eyes."

Last year, Tuvalu's government attracted the world's attention by declaring it would start evacuating its citizens in the face of climate change and rising sea levels. When Australia rejected their proposal for special immigration status, Tuvalu began to negotiate a deal with New Zealand whereby a number of its citizens would be accepted each year, effectively as "environmental refugees". Tuvalu only has a population of around 10,000, and the arrangement is projected to last for 30 to 50 years.

In response to criticism that they are admitting defeat by taking the precautionary approach and planning for gradual relocation, Paani Laupepa, who represents Tuvalu at international climate negotiations, responds, "We have to plan for the future, any responsible government would do that. They are finalizing the number and exact criteria for people to go and when it will start... New Zealand has agreed to our scheme but it is under negotiation."

It is easy for urban-dwelling people in developed countries to underestimate the importance of land to Pacific islanders, and hence the deep personal and cultural significance of its loss. One woman from Kiribati explains, "We can't just move to another country. I would love to go to Fiji. But there I have no land. There I am no one."

The spectre of wholesale relocation raises challenging questions. Once land has been lost, will a residual nationality be able to persist, or does there need to be a new category of "world citizen"? Could such a status be created in acknowledgement of the fact that climate change is a collective problem and requires a collective solution? One of the greatest obstacles in preparing for potential mass movements of environmental refugees is the lack of flexibility in the designation of refugee status. Refugee law needs revising to cope with the problem.

In the event of full-scale national evacuation, there are as yet no plans to deal with an abandoned country's exclusive economic zone, its territorial waters and nationhood. Paani Laupepa expresses a desire to preserve national integrity, even if it means doing it somewhere else, "I think we could have a state within a state, that has to be negotiated with other governments... It will enable us to preserve our sovereign integrity."

Few things could be more sensitive than carving out new territory to create space for a nation. A process that carries a stamp of legitimacy from the UN General Assembly will be needed to handle such a challenge. Poni Favae from Tuvalu's environment department picks up the point, "The UN has the IPCC [to discuss climate change] but it has no equivalent panel to talk about migration." ■

chapter 4

of extreme weather-related events will increase, their intensity will increase, and as these events increase, the countries will really be in a cycle that they can't break out of".

Pacific island nations are on the front line of confrontation with climate change – a change which far from being scientific theory is now described by UN experts as "inevitable".

Atolls like Tuvalu reveal the threats in microcosm. Their fate is echoed around the region and among vulnerable communities across the world. "Our early experience with the real consequences of global warming should be the canary in the coal mine," says Leo A. Falcam, president of the Federated States of Micronesia.

This chapter will focus on the immediate threats to Pacific island nations posed by climate change, and the ways in which conventional development may be fuelling rather than fending off vulnerability. It will examine the options for disaster preparedness and "adaptation" at community, national and regional levels. It will argue that all development decisions from now on have to be viewed through the lens of risk reduction. And it will conclude that far more resources and political will are needed to make meaningful progress in protecting Pacific islanders, and coastal dwellers worldwide, from the worst of the weather.

## Rising seas only part of the problem

According to the World Meteorological Organization (WMO), 2001 was the second warmest year on record. Since 1976, the global average temperature "has risen at a rate approximately three times faster than the century-scale average". In 2001, the Intergovernmental Panel on Climate Change (IPCC), the group of scientists that advise international climate negotiations, produced their Third Assessment Report (TAR). This projects that over the period 1990-2100, global average surface temperatures will climb at a rate without precedent during the last 10,000 years.

Sea levels are projected to rise between 9 and 88 centimetres over the same period. According to the IPCC, "although there will be regional variation in the signal, it is projected that sea level will rise by as much as 5mm per year over the next 100 years as a result of greenhouse gas-induced global warming". The panel says that this rise is "two to four times greater than the rate experienced in the previous 100 years," and adds: "There is new and stronger evidence that most of the warming observed over the last 50 years is attributable to human activities."

The scientists say that "sea-level rise poses by far the greatest threat to small island states relative to other countries. Although the severity of the threat will vary from island to island, it is projected that beach erosion and coastal land loss, inundation, flooding, and salinisation of coastal aquifers and soils will be widespread". These scenarios do not, however, take into account the possibility of sea-level rises due to melting of the Antarctic and Greenland ice sheets. Should seas rise by a metre or more, the consequences are almost unimaginable (see Box 4.2).

Low-lying atolls are not the only islands threatened by sea-level rise, and the accompanying threats of coastal erosion and saltwater intrusion. The vulnerable shorelines of many higher islands and coastal communities are home to large population settlements and critical infrastructure. Small rises in sea levels disguise much bigger effects, such as fluctuating tides and much higher storm surges. According to WMO, "sea-level rise would increase the impact of tropical cyclones and other storms that drive storm surges. The effects would be disastrous on small island states and other low-lying developing countries, such as the Maldives, Tuvalu, Kiribati and Vanuatu where 90 per cent of the population lives along the coasts".

Coastal flooding not only erodes beaches and threatens infrastructure, it inundates soils and fresh water supplies with salt. Many Pacific atolls, such as Tuvalu and Kiribati, are entirely dependent on rain-fed drinking water sources or thin underground freshwater "lenses". However, the combination of changing rainfall patterns and rising seas is threatening this crucial resource. On Tuvalu, there has been an overall decrease in rainfall over the past 50 years. Droughts can last for three or four months, but become a problem after only two or three weeks. On most of the islands the groundwater is not drinkable, according to the environment ministry. And there is not enough capacity to collect and store rainwater.

Salinization of soils and sources of drinking water by higher seas and storm surges may force some islanders to abandon their homes long before the seas swallow them up. Late last year in Vanuatu, one chief on an outlying island recommended that his people relocate to a larger island after saltwater began inundating their land. In the Marshall Islands, farmers are resorting to growing crops in old oil drums, since the islands' soil is now too saline to plant in.

On the Carteret atolls off the coast of Papua New Guinea, rising seas have cut one island in half, and increased soil salt levels, killing off banana and vegetable crops. According to the atolls' district manager, islanders can no longer plant food crops because when they dig they find salt water. Since August 2001, the atolls' 1,500 inhabitants have been surviving on a basic food ration of rice and sweet potatoes shipped in from the mainland. While relocation of the islanders would appear the only long-term solution, the Papuan government lacks the money to move them.

Hilia Vavae is director of Tuvalu's meteorological office, and has worked there since 1981. Her expertise marks her out as special. Hilia was approached in the street by a

chapter 4

woman she didn't know who noticed a bandage on her foot. "Get it mended," said the unknown woman, "you are the most important person in Tuvalu."

Because of the sheer complexity of the science involved, Vavae says that predicting the effects of climate change is an uncertain business. But pointing to 50 years of observations, she says that air temperatures have generally warmed throughout the period. And, she adds, "In the 1980s, only in February did the low-lying areas get flooded, and not very much. But since the late 1990s, and especially now, the frequency of flooding has tremendously increased. Last year we were flooded in November, December, January, February and March – it's quite unusual."

The case of Tuvalu has aroused intense debate about the extent to which sea levels are rising around the atolls. Some sources suggest that tectonic plate movements are also a factor in the atolls' position relative to the sea. In order to clarify what is happening now to absolute sea levels, taking account of land movements, a team of scientists from GeoScience Australia began setting up a series of GPS (Global Positioning System) stations around the Pacific region during late 2001.

Sea-level rise, however, is only one of a range of global warming-induced changes which threaten coastal communities. For example, a rise in sea surface temperatures poses a serious threat to coral reefs. Reefs maintain natural sea defences, supply beach sand and provide habitats for marine animals and fish essential to the local diet. Coral

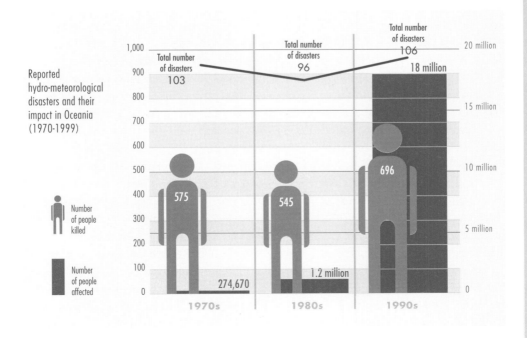

Figure 4.1.

Source: CRED.

## Box 4.2 Treading water: climate change and sea-level rise

In 2001, the IPCC projected that, within a century, global warming could raise sea-levels by up to 88 centimetres. One billion people live at sea level or just a few metres above. Of the world's 19 mega-cities (agglomerations with more than 10 million inhabitants), 16 are situated on coastlines and all but four are in the developing world. But a comprehensive global assessment of the numbers who would be displaced by a one metre rise in sea levels, or even a half metre rise, has not been made. The most vulnerable areas are found in the tropics, especially the west coast of Africa, south Asia and south-east Asia, and low-lying coral atolls in the Pacific and Indian Oceans. The nations hardest hit will be those least able to afford coastal protection measures and where inhabitants have nowhere else to go.

According to a 1998 report by the IPCC, *Regional Impacts of Climate Change*, a one metre rise in sea level would inundate 3 million hectares in Bangladesh, displacing 15 to 20 million people. Viet Nam could lose 500,000 hectares of land in the Red River Delta and another 2 million hectares in the Mekong Delta, displacing roughly 10 million people.

A one-metre rise would swamp about 85 per cent of the Maldives' main island, which contains the capital Male. It could turn most of the Maldives into sandbars, forcing 300,000 people to flee to India or Sri Lanka. "We would have no choice," said President Gayoom as long ago as 1989, "for the Maldives would cease to exist as a nation."

West Africa is at high risk. Up to 70 per cent of the Nigerian coast would be inundated by a one-metre rise, affecting more than 2.7 million hectares and pushing some beaches three kilometres inland. Close to 4 million people would be displaced. Oil production in the Niger Delta could lose US$ 6-18 billion a year. Gambia's capital, Banjul, would be entirely submerged.

In the Mediterranean, Egypt would lose at least 2 million hectares of land in the fertile Nile Delta displacing 8-10 million people, including nearly the entire population of Alexandria. The demise of this historic city would cost the country over US$ 32 billion, close to one-third of annual gross national product (GNP) in 1999.

South American cities would suffer some of the worst economic effects. A one-metre rise in sea level would displace 600,000 people in Guyana – 80 per cent of the population – and cost US$ 4 billion, or 1,000 per cent of its tiny GNP. Meanwhile, the United States has calculated that the cumulative costs of a 50-centimetre rise by 2100 could top US$ 200 billion, not including the huge extra costs of extreme weather events.

As seas rise, coastal land in some regions is sinking. Most large coastal cities have no plans to deal with this. Manila, Bangkok, Shanghai, Dhaka and Jakarta are subsiding, due to the development pressures of excessive groundwater pumping, coupled to urban sprawl (as areas become more built-up, less water seeps into groundwater reservoirs). Saltwater intrusion into coastal freshwater aquifers is a serious problem, compounded by sea-level rise. The cost of new desalination plants, flood-control systems and pumping stations could run into billions.

In Bangkok, rising sea levels would cost an additional US$ 20 million per year in pumping costs alone. Costs for relocating displaced squatter communities would be astronomical. In Shanghai, up to a third of the city's 17 million inhabitants would be flooded, displacing up to 6 million people. Singapore, one city with a comprehensive planning culture, has nothing in its latest 50-year master plan to deal with a one-metre sea-level rise.

"We are overwhelmed right now," shuddered one of Manila's water managers, "I can't even imagine what would happen if the sea rises by a metre. Hundreds would drown during the rainy season and we would be faced with massive capital investments in new, bigger pumping stations and storm-drain systems."

If municipal governments don't begin to take sea-level rise seriously, the term "treading water" is likely to become much more than a metaphor for policy inaction. ■

reefs also provide badly needed foreign exchange earnings through tourism for many small islands. But reef-building corals die when temperatures rise beyond a narrow band.

Coral "bleaching" has already occurred over the past 20 years when sea surface temperatures warmed just one degree Celsius higher than the summer maximum. During the last El Niño in 1997-98, warmer seas affected up to 90 per cent of live reefs around some islands. Such bleaching could soon become a frequent event. According to the IPCC, "the thermal tolerance of reef-building corals will be exceeded within the next few decades". The death of reefs will increase the vulnerability of low-lying shores to rising seas and storm surges.

## Cyclones, droughts and disease will worsen

While changes in sea level and temperature may seem tiny to the untrained observer, they will trigger unpredictable changes in the frequency and intensity of extreme weather events. The latest data on the Oceania region (Pacific islands plus Australia, New Zealand and Papua New Guinea) from the Brussels-based Centre for Research on the Epidemiology of Disasters (CRED) show that these changes may already be under way.

While the overall number of weather-related disasters reported in Oceania has remained fairly constant between the 1970s and 1990s, the impacts of these disasters have become far heavier (see Figure 4.1). Total numbers reported killed by weather-related disasters rose 21 per cent between the 1970s and 1990s. But this rise is dwarfed by an astonishing increase in the numbers of people reported *affected* by such disasters: from 275,000 in the 1970s, to 1.2 million in the 1980s to 18 million in the 1990s – a 65-fold increase. Cyclones affected 18 times more people in the 1990s than in the 1970s, while floods and landslides affected nine times more.

The greatest increase, however, is in the reported impact on Oceania of droughts and extreme temperatures. During the 1970s and 1980s, just one fatality was reported and a total of 71,000 were affected. But in the 1990s alone, drought and extreme temperatures claimed 121 lives and affected over 13 million people. Taken together, these statistics show that the reported impacts of each disaster have escalated dramatically since the 1970s. This suggests that weather-related disasters are becoming more extreme, or that people living in the region are less well prepared and protected than before.

Scientists believe that climate change could make cyclone seasons increasingly unpredictable. Local people, used to more regular cyclone patterns, may in future be less able to prepare. A single cyclone, Kina, hit Fiji in 1993 and caused the redeployment of 32 per cent of the government's budget to pay for urgent reconstruction. In the north-west Pacific, Typhoon Paka hit Guam in 1997 with sustained winds of over 250 kilometres per hour, inflicting more than US$ 600 million of damage.

Normal life comes to a standstill when extreme weather events descend upon small nations (see Box 4.3). Tropical Cyclone Trina struck the Cook Islands in early December 2001. Heavy rains and strong winds hit with very little warning. "I have never seen the seas as high as they were then," said Niki Rattle of the Red Cross, who has lived on the Cook Islands for 29 years. "Many low-lying areas were flooded. The supply of local fruit and vegetables was ruined." International flights stopped, roads were ripped up, Christmas festivities and first-aid workshops were cancelled. On the island of Mangaia local people said flooding was the worst since 1977.

During the past 30 years, tropical cyclones have generally developed over the warmer waters of the western tropical Pacific. During El Niño events, sea temperatures increase towards the central and eastern Pacific, bringing with them more cyclone threats. Scenarios developed by the Commonwealth Scientific and Industrial Research Organisation (CSIRO) suggest that "under climate change, there is likely to be a more El Niño-like mean state over the Pacific". This will increase the threat posed by tropical cyclones to islands in the central and eastern Pacific. The CSIRO research adds that, as atmospheric carbon dioxide levels increase, the intensity of tropical cyclones will also increase, bringing wind speeds 10-20 per cent higher than previously.

More permanent El Niño-type conditions could bring more drought, particularly to the islands of the south-west Pacific. During the intense El Niño of 1997-98, Papua New Guinea was hit by its worst drought of the century. Cloudless skies brought severe frost at higher altitudes, destroying a year of crops in some areas and exposing up to 650,000 people to starvation or disease. According to the South Pacific Applied Geoscience Commission (SOPAC), the drought caused an estimated US$ 100 million of damage. The same El Niño also hit Fiji and the Solomon Islands hard.

Drought slashed Fiji's vital sugar cane crop in half during 1997-98 and caused US$ 18 million in agricultural losses. During the 1982-83 El Niño, rainfall across much of the western Pacific was 70-90 per cent below average. In March 2002, there were early indications that another El Niño is on its way.

Increasingly unstable weather patterns associated with global warming are already having negative health consequences across the Pacific region. "Many tropical islands are now experiencing high incidences of vector- and water-borne diseases that are attributed to changes in temperature and rainfall regimes," says the IPCC. Outbreaks of dengue fever are becoming more common. Malaria, which was previously found mainly in the western and central Pacific, is now extending as far east as Fiji. The disease is also being found at altitudes never seen before in Papua New Guinea.

## Options for mitigation and adaptation

The island nations of the Pacific are diverse in many ways, but they share some common vulnerabilities which hamper their ability to mitigate and adapt to the negative effects of climate change:

- Small physical size and (often) low elevation.
- Wide geographic distribution and remoteness.
- Proneness to "natural" disasters.

In some Pacific islands, the local population can no longer grow their staple crops due to increased salt levels in the soil.

Jerry Galea/
International Federation,
Papua New Guinea.

## Box 4.3 Cyclone strafes Manihiki atoll

*"I'm a big strong man and the way those waves threw me about in the house, I was sure that if it kept up, I wouldn't have been able to fight anymore. If those breadfruit trees hadn't wrapped themselves around the house stopping some of the force of the waves, I don't think I'd be here today. That's a scary experience I'll never forget!"*

*"We thought that we were going to be safe in a two-storey house but the waves kept coming higher and higher and finally when we did come out of the house to run to the highest point, my husband had to hold our brand-new granddaughter above his shoulders as the water around our homes had risen that high! It was a struggle fighting through the water and debris everywhere including trees and houses falling all around you."*

*"One lady and her family took heed of old experiences, took some of their emergency supplies, tied their boat to a coconut tree and rode out the cyclone all night. Some others, who went missing for a few days, had floated to a sister island 25 miles away. They survived by hanging on to an upturned boat. Much later they heard noises from under the boat and found a young girl who was able to breathe in the space between the sea and the bottom of the boat."*

These reminiscences graphically depict the effects of the most devastating disaster to strafe Manihiki atoll in half a century. On 1 November 1997, Tropical Cyclone Martin hit the small coral atoll, killing nine people, leaving ten more missing and destroying over half the island's homes.

Out of a population of 600 people, more than half were airlifted to the mainland of the Cook Islands. There were many injuries from flying iron roofing and debris, exposure to sea water overnight, exhaustion and shock. The island's two villages looked like they'd been bombed. The lagoon between them was full of houses, trees and debris. Many of the dead bodies were found on the beaches. Islanders were banned from eating fish and seafood from the lagoon as a precaution. Some people, especially those with homes that could be salvaged, decided to remain on the island and try to make a go of things. The rest simply left.

The New Zealand government subsequently built two large community cyclone shelters on the island, plus 74 family-sized micro-shelters. The one-room shelters are raised on 8-metre poles, with a 25,000-litre water tank as the ground floor. While the cost was subsidized, each family had to pay US$ 5,700 for their shelter. The Cook Islands Red Cross has visited Manihiki twice since the cyclone and conducted first aid and disaster preparedness training, and formed village disaster groups to improve the coordination of assistance during cyclones.

The lifestyle of Manihikians has totally changed since Martin, as people have relocated from one side of the island to the other, where food was grown and livestock were reared before the cyclone hit. The lucrative black pearl industry had been the main factor in keeping people on the island. But even black pearls could not tempt back many of those who evacuated. ■

- Rapid urbanization and dense, growing populations.
- Increasing degradation of fragile environments.
- Limited natural, human and financial resources.
- Loss of traditional coping mechanisms.
- Export-dependent, open economies.

The IPCC says that the overall vulnerability of small island states is "a function of the *degree of exposure* of these states to climate change and their *limited capacity* to adapt to projected impacts". Options for mitigation and adaptation available to richer, larger nations are simply not possible for small island states.

For example, in the case of sea-level rise and coastal erosion, a wealthy island nation such as Singapore is likely to invest in structural protection (beach fills and break-waters) for developed coastal areas and tidal gates to prevent major canals from sea-flooding. However, this level of investment is often not financially feasible for cash-strapped Pacific islands. Sometimes, hard engineering solutions have been poorly planned, increasing coastal vulnerability. According to Kakee Kaitu, a senior assistant to the prime minister of Tuvalu, "Sea walls don't solve anything. We tried them to stop coastal erosion, now you go back and the walls are gone."

Softer alternatives to protect coastlines include enhancing natural protection by replanting mangroves (see Box 4.4) and protecting coral reefs. Artificial nourishment (heightening beaches with shipped-in sand or aggregate) is a more costly option. The IPCC has even suggested that nations which lack adequate materials for artificial nourishment could consider pillaging sand from less important islands in order to build up the levels of more important islands.

If beach nourishment proves too costly, another alternative is "managed retreat". This involves prohibiting building projects near the shoreline, enhancing the natural resilience of coastlines by encouraging dunes, lagoons and estuaries to grow, and allowing the coastline to "recede to a new line of defence". Building codes could even stipulate that houses should be built on stilts.

However, for the lowest-lying Pacific atolls, managed retreat may not be a viable option, as there is nowhere to retreat to. Most critical infrastructure such as social services, airports, port facilities, roads, tourist developments and other utilities are within 100 metres of the shoreline. In such cases, hard engineering protection or mass migration may be the only alternatives. Some theoretical work has been done in the region on making buildings disaster-resilient. Tuvalu's climate change officer Seluka Seluka points out that a flood- and cyclone-resistant house can be built for around US$ 16,000. But he questions whether this level of spending is sensible if the land the houses are built on may soon be lost to the sea.

Sound water management is a key mitigation policy, especially for Pacific islands where rainfall is predicted to become more variable and saltwater is intruding into drinking water supplies. Water will have to be carefully allocated, recycled and possibly even rationed. Rainwater will have to be more efficiently harvested – perhaps Pacific islands could learn age-old harvesting techniques from other drought-stricken nations such as Sri Lanka (see Chapter 1, Box 1.2). Desalination is a very expensive alternative.

Diseases such as malaria and dengue will need combating with more preventive health care, health education and early warning. The spread of deadly disease-carrying mosquitoes can be reduced by draining swampy areas and distributing insecticide-proofed bednets. The IPCC suggests that very simple technology, such as using sari cloth to filter drinking water, could reduce the risk of transmitting water-borne diseases like cholera.

Meanwhile, early warning systems will be more important than ever as extreme weather events become harder to predict under climate change conditions. The United States National Weather Service monitors meteorological conditions across its territories in the north-west Pacific and can give advance warning of extreme events. And WMO has set up a system of cyclone early warning for the south Pacific.

## Adaptation initiatives lack resources

Initiatives at various levels are under way to combat the risks that climate change poses to Pacific island nations. But more resources and political urgency are needed.

At the international level, progress on the world's only binding agreement to tackle global warming, the Kyoto Protocol, has been hampered by the United States' refusal to ratify. But this is by no means the only concern. In 2001, the IPCC stated, "analysis has shown that even with a fully implemented Kyoto Protocol, by 2050 warming would be only about 1/20th of a degree less than what is projected by the IPCC. Therefore, climate change impacts are inevitable."

Given that the impacts we have discussed are inevitable, exposed nations have no choice but to adapt to these new risks. Little is know about the costs of adaptation: scientists and policy-makers put the worldwide cost at anything from tens to hundreds of billions of dollars per year. Last year, however, rich countries pledged to provide just US$ 0.4 billion per year by 2005 to help developing countries adapt to climate change. By contrast, industrialized nations spend US$ 70-80 billion per year on energy subsidies, including for fossil fuels (see Box 4.5).

Various regional initiatives sprang up during the 1990s, designated the International Decade for Natural Disaster Reduction (IDNDR), including the Pacific Islands

## Box 4.4 Mangrove planting saves lives and money in Viet Nam

Viet Nam is one of the most typhoon-lashed nations in Asia. Every year, an average of four sea-borne typhoons and many more storms wreak havoc on this low-lying country. In what may seem a curious pursuit for a humanitarian organization, the Viet Nam Red Cross (VNRC) has, since 1994, been planting and protecting mangrove forests in northern Viet Nam.

The reason for its commitment to mangrove protection, which has included planting nearly 12,000 hectares of trees and defending them from shrimp farmers who want to hack them down, is a simple one: mangroves protect Viet Nam's coastal inhabitants from the ravages of typhoons and storms. These submerged, coastal forests act as buffers against the sea, reducing potentially devastating 1.5-metre waves into harmless, centimetre-high ripples.

The mangroves planted by the Red Cross protect 110 kilometres of the 3,000-kilometre sea dyke system that runs up and down Viet Nam's coastline. The VNRC, with financial support from the Japanese and Danish Red Cross, plant four different species, which reach a height of 1.5m after three years.

The benefits are staggering. In financial terms alone, the mangrove programme proves that disaster preparedness pays. The planning and protection of 12,000 hectares of mangroves has cost around US$ 1.1 million, but has helped reduce the cost of dyke maintenance by US$ 7.3 million per year.

In lives spared, one need only look to the dividend reaped during Typhoon Wukong in October 2000. This typhoon pummeled three northern provinces, but left no damage to the dykes behind regenerated mangroves, and no deaths inland of the protected dykes. In the past, waves would breach the coastal dykes and flood the land of poor coastal families.

As well as the lives, possessions and property saved from floods, the VNRC estimates that the livelihoods of 7,750 families have benefited from the replanting and protection of the mangrove forests. Family members can now earn additional income selling the crabs, shrimp and molluscs which mangrove forests harbour – as well as supplementing their diet.

The presence of additional mangrove forests in a country that has seen them decimated in the last 50 years, due to shrimp farming, coastal development and chemical defoliants dropped during the Viet Nam war, is crucial. As sea temperatures and levels rise, more severe typhoons and storm surges can be expected. This could be disastrous for the inhabitants of Viet Nam's east-facing coastline.

This risk has spurred the Red Cross to continue its investment in mangrove regeneration, despite continued threats from coastal shrimp farmers and developers. It is just as well. Those who live inland of sea dykes are a little less at the mercy of typhoons and storms now. And they hope to keep it that way. ■

Climate Change Assistance Programme (PICCAP). PICCAP aims to build the capacity of Pacific nations to plan for and adapt to climate change, by training individual country teams to lead the process.

Accompanying IDNDR's work in the 1990s was a regional process called the South Pacific Disaster Reduction Programme. This left in place disaster management offices and plans at the national level across the region. An array of conferences, publications, improvements in technical understanding, access to scientific information and an appreciation of the importance of community-level organization are all legacies of the programmes. So too is the growing international awareness of the particular vulnerability of small island states.

But national disaster preparedness arrangements have attracted criticism from a number of agencies working in the field. For example, disaster management officials are often temporary political appointments working with civil servants whose jobs often change; these factors conspire to destroy institutional memory. Villagers often have as good an understanding of the risks and challenges as scientists and officials, yet they are not listened to. There is a lack of clarity about who "owns" disaster management plans. Too often these plans refer only to preparedness for disaster response and fail to address the full range of risk reduction priorities. Taking responsibility remains a problem. In Fiji, for example, poor coordination and confusion defining "natural" versus man-made disasters means that, between the police, fire department and army, "no one is sure whose job it is to do what," says the UN's Charlie Higgins. And in smaller nations like Tuvalu, so much time and money are taken simply running the machinery of government, that few resources remain to deal with disasters or threats from climate change.

Higgins adds, "Most plans emerged from standard models. They didn't emerge from the particular circumstances of the country. And, therefore, they don't guide appropriate actions." His concerns are echoed by Seo Seung Chul, the International Federation's regional disaster preparedness coordinator. "Many government plans are unworkable. They get written by [people in] Australia or New Zealand – not by people on the ground. The plans are not understood or shared among government departments. They're good on paper but most wouldn't work." Chul sees that a decade of raising awareness of disasters and laying plans to deal with them has missed out one vital ingredient. "The governments have plans, but when you ask them what their budget is for disaster preparedness, the answer is 'zero'."

Institutional support for national disaster management offices in the South Pacific is provided by the technical agency SOPAC, whose scientific support to government officials is of undeniable value. One of SOPAC's proudest achievements is the development of sophisticated computer models using highly detailed computer-generated graphics that show, for example, the precise anticipated impact of a tsunami hitting a coastal city. With this, they aim to win arguments with planning officials to take risks more into account in daily decision-making. SOPAC also hopes the models could be an effective visual tool to persuade communities to prepare better, by showing exactly what would happen to their homes in the event of serious flooding.

However, other agencies are less convinced that occasional visits by consultants with laptop computers will be taken seriously at the community level, or even understood. Though potentially effective with decision-makers at a senior government level, in terms of essential community work, it appears to be a top-down preparedness product in search of a market.

Public awareness of how to prepare for disasters can be boosted through cheaper and less technical means. For example, the national disaster management office in Fiji produces a poster and office calendar, which proclaim, "Disasters Do Happen. ARE YOU PREPARED?" There follows a series of cartoon pictures, accompanied by simple words of advice: "TSUNAMI – RUN", "EARTHQUAKE – DUCK", "FIRE – DON'T PANIC" and "FLOOD – EVACUATE".

## Community-based preparation

Because official assistance tends to concentrate overwhelmingly on disaster response, even when warnings are available, people are usually "on their own" in the preparation phase. Because of the remoteness of many communities they are often left that way for days afterwards as well. In an atoll nation like Tuvalu, the outer islands are up to three days' sea journey from the main administrative centre, Funafuti. In case of disaster, the government has just a single boat (and no aircraft) to provide whatever assistance is available.

Interestingly, the IPCC notes that "a key misconception is that adaptation is a task carried out by governments… Most adaptation, however, will be carried out by individual stakeholders and communities, urban or rural, that inhabit island countries. Therefore, the government's primary role is to facilitate and steer this process". The IPCC goes on to argue that "because strong social and kinship ties exist in many small island states – for example, in the Pacific – a community-based approach to adaptation could be vital if adaptation policies and options are to be successfully pursued".

So, does community-based action provide the answer to arresting the rising tide of climate change-related hazards? Roshni Chand, project manager at the Foundation for the Peoples of the South Pacific (FSP) based in Suva, Fiji, certainly believes so. She concedes that much of the high-level technical and policy work is necessary, but adds, "It can be meaningless to local people. They are not disaster managers." FSP works to increase cyclone preparedness and reduce vulnerability at the village level by training community groups. They have produced a guide on training trainers to multiply local disaster awareness. They emphasize that community-based disaster preparedness is about values and attitudes rather than learning facts. "It's about taking responsibility for our own families and possessions in practical ways," says Chand. FSP stresses that the skills this approach concentrates on – leadership,

## Box 4.5 Fossil fuel subsidies outstrip adaptation funds

The sums for adaptation to climate change have not been done, but working them out is now a priority. Without even a general target, it is impossible to know how far the international community's pledges of support fall short.

The secretariat of the 1992 Earth Summit produced one of the only related estimates. They said implementing the Agenda 21 plan for sustainable development in low-income countries would need an extra US$ 125 billion per year from rich countries in the form of aid or other concessions. This analysis was ignored and aid has been in relative decline ever since. More recently, Munich Re's chief geoscientist, Gerhard Berz, has calculated that the projected costs of *damage* inflicted by climate change could top US$ 300 billion per year within the next few decades. One thing is sure: poor countries will need more new resources to mitigate and adapt to the impacts of global warming.

One of the first attempts to value the services provided by ecosystems put a monetary value on them of US$ 33 trillion per year, according to an article by R. Costanza, published in the journal *Nature*. Water and climate regulation alone were valued at US$ 4.1 trillion. Flood and storm protection were valued at US$ 1.1 trillion, and coral reefs, which are particularly vulnerable to climate change, provided US$ 375 billion worth of goods and services. Such environmental services should be worth real investment to protect.

Fossil fuels provide 87 per cent of global commercial primary energy, according to the International Atomic Energy Authority. But burning them also fuels climate change. They power the global economy, bringing disproportionate wealth to industrialized countries. Yet the world's poorest countries are twice as efficient as the richest in turning fossil fuels into national income.

Conservative estimates suggest that the Organisation for Economic Co-operation and Development (OECD) group of rich countries spends around US$ 70-80 billion per year on energy subsidies, including fossil fuels and fossil fuel-based activities. The International Energy Agency estimated that the German coal sector alone received US$ 10 billion in subsidies in 1996. Yet while industrialized countries find significant sums to subsidize their own use of climate-change-causing fossil fuels, only a trickle has been made available to clear up the mess they cause in developing countries.

The main conduit of funds dedicated to sustainable development is the Global Environment Facility (GEF) operated jointly by the World Bank and UN. It administers three new funds under the Climate Convention and Kyoto Protocol: a special climate change fund, a least developed countries fund and an adaptation fund. In the year 1999-2000, GEF funding for climate change was US$ 1.4 billion. Only US$ 199 million of this was grant funding.

At climate talks in Bonn last year, rich countries pledged to provide US$ 0.4 billion per year by 2005 to help developing countries "manage their emissions and adapt to climate change". That figure is less than 1 per cent of the amount that industrialized countries spend subsidizing their own use of fossil fuels. ■

planning, responsibility, cooperation and commitment – are all inherently useful and transferable to daily life.

Trainees are expected to be able to spot factors that increase general vulnerability and to identify vulnerable people. They help develop contingency plans, map local capacities, encourage local people and institutions to take on disaster preparedness roles, and explain the value of simulating disasters in order to practise responding to their aftermath. Everything from communication to counselling and first aid is included.

But Chand laments that even their well-thought-out community work, which has been tried and adapted in Palau and Vanuatu, is patchy and poorly supported by government authorities. "Nothing really happens. If you go to the national disaster management office there are very few staff," she says. "They are not able to reach the community. And donors and international NGOs and all the larger bodies are not doing much to help to take all the mitigation and preparedness initiatives down to the community." Not only is most disaster preparedness top down, Chand believes, "but then it stops before it gets to the community".

National Red Cross societies may offer one way of bridging the gap, since they are both rooted in the community and legally established as "auxiliaries" to the government in times of crisis. The Red Cross approach to mitigation and preparedness is built on community-based self-reliance (CBSR). The key to the approach is the vulnerability and capacity assessment (VCA) which, carried out by local people, seeks to identify the factors that both increase and decrease the community's exposure to risk. Simply going through the VCA process often improves community disaster awareness and strengthens local resolve to do more to reduce risks (see Chapter 6, Box 6.4).

As part of the assessment, participants are encouraged to draw maps of their local community. They identify vulnerable places such as bridges and houses on steep slopes; vulnerable people such as the elderly and disabled; and potentially hazardous locations like latrines, deep-water lagoons or even areas of long grass hiding snakes. They also map resources such as strong buildings to use as evacuation centres.

Red Cross staff then train teams of local volunteers through community workshops. To ensure that the training manual reflects the "priorities, customs and beliefs" of Pacific nations, it was tested in Papua New Guinea and Vanuatu and then refined. Islanders are not only trained to set up community disaster preparedness committees and plans, they also receive first-aid training to make them more self-reliant following disasters. To enhance community self-reliance, the Red Cross has established 44 shipping-style containers of emergency supplies at key locations across the region, along with three warehouses of basic relief items.

## Box 4.6 Socially responsible economies key to disaster reduction

History shows that economies planned with social responsibility in mind are a key defence against vulnerability to disasters. Conversely the wrong kind of economic structures can increase the impact of unstable climatic conditions on the planet's poorest.

In *Late Victorian Holocausts: El Niño Famines and the making of the Third World*, author Mike Davies argues that 19th century economic globalization left India, China and Brazil tragically exposed to "natural" disasters. According to Davis, the "forcible incorporation of smallholder production into commodity and financial circuits controlled from overseas" fundamentally undermined food security, and left millions of people exposed to famine during El Niño cycles.

Indian peasants, for example, had three practical safeguards against famine conditions provoked by climate instability: domestic grain hoards; family ornaments (silver); and credit with the village money-lender and grain dealer. Towards the end of the 19th century, all these were lost under British rule due to the changing balance of power in the rural economy and new trade imperatives.

As local economies became violently incorporated into world markets from 1850 onwards, argues Davies, Indian peasants and farm labourers became more exposed to natural disasters. Under the British, "Between 1875 and 1900, years that included the worst famines in Indian history, annual grain exports increased from 3 to 10 million tons – an amount equivalent to the annual nutrition of 25 million people."

Yet the impact of climate-related disasters can be drastically reduced by alternative economic regimes. Before the British took control of much of India, local Mogul rulers used a range of policies to prevent famine taking hold. They relied on embargoes on food exports, anti-speculative price regulation, tax relief and distribution of free food without a forced labour counterpart. And they zealously policed the grain trade in the public interest.

Thirty-one serious famines happened in 120 years of British rule in India, claiming the lives of millions. Only 17 famines were recorded in the previous 2,000 years. For Davies, the experience represents "a baseline for understanding the origins of modern global inequality... how tropical humanity lost so much economic ground to western Europeans after 1850". He adds that it "goes a long way to explaining why famine was able to reap such hecatombs [large death tolls] in El Niño years." ■

## Indigenous knowledge key to resilience

Despite a decade of high-tech and low-tech disaster reduction initiatives, for most islanders little has changed in how they combat natural hazards. When they hear a cyclone warning, for example, they instinctively know what practical preparations are necessary: food, drinking water and shelter are the absolute essentials to secure. Acting under their own initiative, villagers will seal plastic containers and barrels containing food and drinking water. Any tree branches overhanging houses will be cut and coconuts that could fall onto roofs removed. Tall trees are usually allowed only at a safe distance from village housing. Islanders will often sink posts into the ground and erect basic windbreaks, made of matted coconut fronds. They may lash house beams to a combination of posts and tree trunks to add structural support.

Many island societies and economies are traditionally based on cooperation, gift exchange and sharing rather than on personal gain and competition for profit. According to anthropologists Keith and Ann Chambers, "sharing equalizes access to resources across a community". This culture of mutually supportive communities can prove invaluable when disaster strikes. But the intrusion of a competitive and acquisitive economic culture directly affects communal vulnerability, say the Chambers. And profit-seeking enterprise initiatives pushed by aid projects "support the weakening of sharing obligations," they argue.

Mataaio Tekenene, Tuvalu's environment officer, agrees, "In the past people had land, trees, the tools they needed and the knowledge from ancestors. It was all they needed to survive… The shift to a modern lifestyle, a western lifestyle is making people more vulnerable to the climate. They have lost their ways of coping and managing."

Some have sought to catalogue these local strategies before they vanish. A Pacific region conference in December 2001 brought people together to share and codify knowledge of tried and tested indigenous coping mechanisms. Their integration into official disaster management is increasingly seen as a priority. Traditional resource management practices are embedded, for example, in the stories (*mo'olelo*), chants (*mele*) and dances (*hula*) of native Hawaiians.

Old traditions can be seen re-emerging in projects like FSP's community theatre project, "A Dramatic Tool for Cyclone Preparedness", produced in response to the series of cyclones that hit the region in the late 1990s. Seeing the weakness of official and technical preparations, FSP developed a model for community theatre that is fun, participatory, adaptable for different communities and equally applicable to issues ranging from health to the environment. Most importantly it is effective because it is rooted in the way that most islanders learn their history, through "songs, dance, rituals and legends that were handed down from one generation to the next". The plays use audience participation and follow-up discussions. To measure the plays' effectiveness, audiences are invited to fill out evaluation forms to show what they have learned. One theory behind the theatre is that people remember only 20 per cent of the information they hear but 90 per cent of what they hear, see, talk about and do.

Indigenous cultures offer broad frameworks for decision-making in an increasingly unstable world. According to a report published last year by the Hawaii-based East-West Center, old models for effective community participation include the Hawaiian '*aha* council, in which "all affected parties work to resolve resource management concerns through sustained dialogue and shared decision-making". And policy planners of all nations could benefit from the Carolinian principle of *meninkairoir*, which means "taking the long view". Many other specific examples of indigenous knowledge give advice on sustainable water and plant management; preventing beach erosion; predicting storms; and preserving dedicated islets for food in times of scarcity and disaster.

## Development can fuel vulnerability

While the westernization of Pacific societies may be eroding traditional coping measures, Tuvalu's Mataaio Tekenene goes even further, "Both climate change and development are killing the island."

The development pressures of population growth and urbanization are threatening to increase Pacific islands' vulnerability to climate change. Since 1975, Tuvalu's population has soared by 50 per cent, increasing the drift of people from rural areas and outlying islands to urban centres. Lack of decent land and housing has led to an increase in vulnerable squatter settlements. In the Solomon Islands, Papua New Guinea and elsewhere, new homes are being built on plots exposed to flooding and landslides.

Local development priorities are not always maximized to reduce risks. For example, Tuvalu recently spent over US$ 3 million, one-third of its annual budget, on a road-building programme. Despite a housing shortage, only US$ 107,000 was spent on building and renovating homes. The priority seems even stranger considering that the main island is flat, has a speed limit of 25 kilometres per hour, and you could walk from one end to the other in an hour and a half. Tekenene complains, "Things have been done without proper consultation. Now it's too late, nothing can be done. The building company had already arrived before we could raise our concern."

The shift from a traditionally self-sufficient economy to an export-led economy carries risks too (see Box 4.6). Over-dependence on a few key exports exposes island states to the impacts of climate change more than many larger nations. Some islands are very dependent on tourism for a sizeable proportion of their income. Tourism in Samoa accounted for nearly half of all foreign exchange earned by exports of goods and services in 1997. For Vanuatu, the figure was 41 per cent, and for Fiji it was 29 per cent. For countries represented by the South Pacific Tourism Organization, the industry employs between 15-20 per cent of the workforce. Yet the beaches, coral reefs and coastal infrastructure which tourism depends on are seriously threatened by climate change.

Reliance on exports of agricultural commodities adds to Pacific islands' "susceptibility to external shocks" says the UN Development Programme (UNDP). For some island nations, agriculture accounts for 75 per cent of foreign earnings. But where cash crops are cultivated at the expense of subsistence agriculture, countries can fall vulnerable to fluctuations in both world markets and global weather.

During the 1990s, Samoa was hit by two "once in 100 year" cyclones and the loss of its main crop, taro, due to disease. Instead of famine, Samoa recovered. Its resilience to disaster is partly attributed to the traditional food production system, which uses a wide variety of crops, bred over generations to be hardy, and grown together in a robust mixed-crop pattern. The breakdown of such systems across the region is, according to UNDP, causing "repercussions on food security and vulnerability to other disasters".

Peter Waddel-Wood of AusAID, a principal donor for many Pacific states, is aiming to increase food security: "Our focus is on self-reliance and self-sufficiency because of the particular problems of the region." A range of basic crops, with hundreds of varieties, provides the basic diet of South Pacific islanders. These include banana, coconut, breadfruit, taro, *pulaka* (a popular variety of taro), squash, yam, cassava, papaya, sweet potato (470 varieties) and paw paw, plus fish, pork and chicken.

According to Jimmy Rogers, deputy director of the South Pacific Commission, conventional mono-cropping increases vulnerability, "Mono-cropping is OK if you can avoid disasters, but you can't avoid disasters either natural or economic in terms of fluctuating prices." Rogers believes it is essential to train university students in the importance of preserving the huge local range of hardy crop types. In the Solomons, hardy but less sweet varieties of sweet potato were allowed to become extinct.

Both neglect and commercialization are to blame. "Prevention is always better than cure. Commercialization could create pressures to lose many of the old varieties," says Rogers, "We have to think say 50 years down the line when a new pest emerges."

And prevention pays. New Zealand, for example, installed a US$ 1.3 million detection system which picked up the presence of papaya fruit fly. Australia had no similar system, suffered an outbreak of the fly, and spent over US$ 50 million trying to contain it. Experts estimate that a detection system would have cost them around US$ 10 million.

## New development paradigm

The threats posed to Pacific islands by climate change are varied and far-reaching. Unprecedented changes in temperatures, sea levels and weather patterns bring incalculable risks not only to the Pacific's natural environment, but to economic development, health, food security and public safety. Conventional development is too often "disaster blind", it risks exacerbating the vulnerability of island nations and coastal communities to the volatility of world weather.

A new development paradigm is needed, in which risks are proactively assessed, prioritized and reduced. Every policy decision at every level must pass the acid test of whether it will increase or decrease vulnerability to the effects of climate change. From now on, planners have no choice but to view all development decisions through the lens of risk reduction. Crucially, communities at risk must be at the centre of this planning process if it is to succeed.

Specific and urgent priorities include:
- global assessment of the likely costs of adaptation to climate change in poor countries;
- new funds made available by industrialized countries for poor country adaptation that are at least equal to the value of rich country subsidies to their domestic fossil fuel industries;
- development models based on risk reduction and incorporating community-driven coping strategies in adaptation and disaster preparedness;
- disaster awareness campaigns with materials produced at community level and made available in local languages; and
- coordinated plans, from local to international levels, for relocating threatened communities with appropriate political, legal and financial resources.

Small island states cannot achieve this alone. Political and financial commitment from developed nations is needed. The IPCC points out that "small island states account for less than 1 per cent of global greenhouse gas emissions, but are among the most vulnerable of all locations to the potential adverse effects of climate change and sea-level rise". What is happening today to low-lying communities is a warning to countless other regions around the globe. A stronger, realistic commitment from the world's richest, industrialized nations to reduce the risks posed by climate change will prove in their best interests. It is long overdue.

*Andrew Simms, policy director at the New Economics Foundation, and an advisor and writer on environment, development and globalization issues was principal contributor to this chapter and boxes 4.1, 4.5 and 4.6. Box 4.2 was contributed by Don Hinrichsen, a writer on coastal development issues; Box 4.3 by Niki Rattle of the Cook Islands Red Cross; Box 4.4 by Rohan Kay and Ian Wilderspin, respectively regional information officer and regional disaster preparedness delegate at the International Federation's regional delegation in Bangkok.*

## Sources and further information

Barnett, Jon and Adger, Neil. *Climate Dangers and Atoll Countries*. University of Canterbury (New Zealand) and University of East Anglia (UK), October 2001.

Chambers, Keith and Ann. *Unity of Heart: Culture and Change in a Polynesian Atoll Society*. Waveland Press, 2001.

Davies, Mike. *Late Victorian Holocausts: El Niño Famines and the Making of the Third World*. London: Verso, 2000.

de Moor, Andre. *Subsidies and Climate Change*, Natural Resources Forum, May 2001.

de Moor, Andre and Peter Calami. *Subsidizing Unsustainable Development: Undermining the Earth with Public Funds*. Institute for Research on Public Expenditure/Earth Council, 1997.

Depledge, Joanna. *Climate Change in Focus: The IPCC Third Assessment Report*. Briefing Paper No 9, Royal Institute of International Affairs, February 2002.

Intergovernmental Panel on Climate Change (IPCC). *Third Assessment Report*. Geneva: IPCC, 2001.

International Federation of Red Cross and Red Crescent Societies, *World Disasters Report 2001*, Geneva: International Federation, 2001.

Shea, Eileen. *Preparing for a Changing Climate: The Potential Consequences of Climate Variability and Change, Pacific Islands*. East-West Centre, Hawaii, October 2001.

United Nations Development Programme (UNDP). *Pacific Human Development Report*. New York: UNDP, 1999.

World Meteorological Organization (WMO). *Statement on the Status of the Global Climate in 2001*. Geneva: WMO, 2001.

World Resources Institute, adapted from R. Costanza et al, "The value of the World's Ecosytem Services and Natural Capital" in *Nature*, Vol. 387.

## Web sites

East-West Center **http://www.eastwestcenter.org**

GeoScience Australia **http://www.agso.gov.au/**

International Atomic Energy Authority **http://www.iaea.org.at**

Intergovernmental Panel on Climate Change (IPCC) **http://www.ipcc.org**

South Pacific Applied Geoscience Commission (SOPAC) **http://www.sopac.org.fj/**

South Pacific Commission **http://www.spc.org.nc/**

World Meteorological Organization **http://www.wmo.ch**

# chapter 5

## Section One

### Focus on reducing risk

# Reducing earthquake risk in urban Europe

South-eastern Europe is awaiting another earthquake. In Romanian living rooms, the population has been horrified to hear scientists announce on TV that a major earthquake can be expected within the next five years. They have reason to fear. The most serious in recent years was in 1977 when a quake measuring 7.2 on the Richter scale left 1,650 people dead and 10,000 injured. Bucharest was worst affected. Other earthquakes followed in 1986, 1990, 1991 and 1996. A large tremor in March last year panicked the population, but thankfully caused no material damage.

Turks share the fear, and shadows of the recent past impact upon their lives. Where 1999's two earthquakes left up to 20,000 people dead and 50,000 injured in north-western Turkey, newspapers have reported many cases of people jumping from windows in panic when tremors occur. In February this year, a quake measuring 6 on the Richter scale killed over 40 people in central Turkey and damaged the homes of up to 60,000 inhabitants.

Albanians have social crises and grinding poverty to preoccupy them but the chance of seismic disaster is by no means forgotten. The country's last major earthquake was 20 years ago and a big one tends to come around every 20-25 years. Given the country's problems, the humanitarian consequences of a major quake are unthinkable, just as they are in neighbouring Macedonia. A repeat of the 1963 Skopje quake, in which 1,066 people died and much of the town was demolished, would be devastating.

According to data from the Brussels-based Centre for Research in the Epidemiology of Disasters (CRED), earthquakes have proved by far the most deadly of all Europe's disasters over the past decade. From 1992-2001, earthquakes claimed 58 per cent of the 36,000 European lives reported lost due to both "natural" and technological disasters. And quakes cost Europe over US$ 27 billion in estimated damage during the decade, making them the continent's second most expensive type of disaster after floods.

Since collapsing buildings kill most earthquake victims, risk is more concentrated in urban areas. So how are European cities planning to reduce these risks? Enforcing better building codes and strengthening "lifeline" infrastructure, such as hospitals and schools, emerge as key mitigation measures. But this takes time, money and willpower. So what can be done now to prepare for the catastrophe round the corner?

Photo opposite page: Making buildings earthquake-resistant is a crucial long-term strategy. But structural mitigation must be complemented by widespread disaster awareness and preparedness measures, so that exposed communities know what to do when disaster strikes.

Mikkel Ostergaard/ International Federation, Turkey 1999.

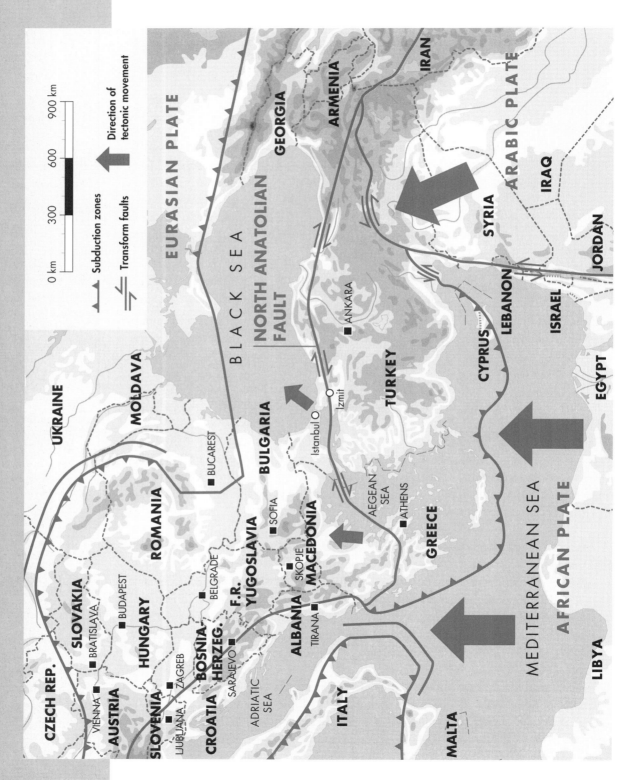

chapter 5

How can structural and non-structural initiatives complement each other in reducing the deadly toll of earthquakes? This chapter will analyse options for:

- **Mitigating earthquake risk.** How to prevent or reduce the risk of an earthquake's disastrous impacts.
- **Preparing for earthquakes.** How authorities and exposed communities can prepare to respond better to disaster.
- **Converting ideas into action.** Advocating for the commitment and resources necessary to make mitigation and preparedness happen.

## Mitigating earthquake risk

The earthquakes which hit the Turkish towns of Izmit and Duzce in 1999, known collectively as the Marmara earthquakes, not only took a terrible human toll. They cost the country around US$ 20 billion in damage alone – equivalent to over 10 per cent of annual gross domestic product (GDP). And that doesn't account for the knock-on effects to the economy. What has been learnt since then, and what has changed? Can Turkey afford not to be ready next time?

The Marmara quakes, with epicentres barely 100 kilometres east of Istanbul, brought widespread criticism of the national state of disaster mitigation and preparedness for response. They changed the way Turkey thought about earthquakes – and certainly focused minds in Istanbul itself. Most experts consider that the city runs a very high risk (between 60 and 70 per cent) of being struck by a major earthquake within the next two to three decades. That may not mean in 30 years. It could mean tomorrow.

Most of Turkey is seismically active territory. Two major fault lines, the south and north Anatolian faults, run the length of the country from east to west. Of the big Turkish cities, only Ankara, the capital, is not seriously earthquake prone. The Marmara quakes were on the north Anatolian fault. Some specialists think the next big one will be further west, almost directly on the southern edge of Istanbul. The palpable sense of urgency makes the city a poignant case study of lessons learned, and what it is possible to do with them.

When precisely a quake might strike is beyond the scientists. Through studying tectonic plates, fault lines and surface geology, they can predict possible disaster locations fairly accurately. And mathematical models can help to predict probabilities of time scale, but not specific times. Sophisticated equipment has made it possible to identify a very small interval (up to 15 seconds) between warning signs and the moment tremors are felt. It could be long enough to activate a direct cut-out of the main gas distribution or electrical substations, which would diminish the risk of

## Box 5.1 "Grandpa Earthquake" dispels fear of disaster

"The worst possible reaction to an earthquake is panic." This is the message of Ahmet Metin Isikara, director of the Kandili Observatory, situated on a hill high above Istanbul. For thousands of children and their parents, Isikara has become "Grandpa Earthquake". In its campaign against fear, the Kandili Project began with teachers. At least one teacher from each of the city's 3,000 schools has been trained to train other teachers and, ultimately, the children in how to prepare for the day when disaster strikes.

Isikara stars in a series of short animated films which show children and adults exactly how to react in an earthquake. With his shock of white hair and toothbrush moustache, Isikara is stopped on the street by children saying things like, "Hello there, Grandpa Earthquake. Like you told me, I'm not afraid anymore." His blue eyes twinkle and he obviously relishes his star quality. But he turns absolutely serious when he talks about the importance of preparing a population to deal with earthquakes, and of going through the children to reach whole families.

The films are skilfully made and really reflect Turkish culture: the homes where children help Grandpa bolt down the furniture, for example, look like real Turkish interiors, not those of North American soap operas. The enthusiasm at Kandili is infectious. Disaster training has been incorporated into the curriculum of grades 1 to 8 and schools hold disaster preparedness days. Future plans include creating an "earthquake park", an information centre with explanatory exhibits, and even a simulated earthquake room so that children can practice "for real" what they have been taught in theory.

The message is beginning to get across: "It doesn't matter how big it is; what counts is *Are you ready?*" Turkish children are disciplined and good disseminators of information. Earthquake survival kits – including items such as bottled water, torch, radio, photocopies of the family's important papers – are appearing in homes and even offices. However, admits Isikara, the terrible thing with "earthquake education" is that there is no way to tell how well it works until the next one hits.

Isikara, who is well connected within the Turkish establishment, wants to go beyond what he feels are sterile academic quarrels. He lectures incessantly and appeals to a broad cross-section of society. He reminds politicians and economic leaders that Turkey lost around US$ 20 billion because of the 1999 earthquakes and that being prepared means losing less next time. To him, and other astute observers, those earthquakes triggered the serious economic crisis in which the country was still embroiled two years later. If the earthquake had hit the country's economic heart, Istanbul, the situation would have been far worse. ■

fires. But it is still insufficient to provide the population with enough warning to evacuate buildings.

The sense of urgency is even greater since, in recent decades, Istanbul has grown almost exponentially – today the population is well over 10 million. Some parts of the city reach tremendous human density, with more than 40,000 people per square kilometre, increasing the risk of disaster. Some experts believe that, depending on the location and magnitude of a future quake, 30 per cent of Istanbul's 900,000 buildings could collapse completely. With a large proportion of the population living in flats, typically five- or six-storey buildings squashed together along steep and narrow streets, the consequences would be catastrophic.

## Substandard construction costs lives

The Marmara quakes shook up more than the earth in Turkey, and left no illusions. In the first place, the number of deaths would have been dramatically less if the country had enforced its building regulations. It is a common shortcoming in south-eastern Europe and more tragedy will undoubtedly occur elsewhere because of it. Mihail Garevski, from Macedonia's Institute of Earthquake Engineering and Engineering Seismology (IZIIS) in Skopje, was the leader of an expert team that visited Izmit a few days after the 1999 disaster. He is unequivocal: "Failure and damage of structures is inevitable in catastrophic earthquakes. However, such an extensive failure as that in Izmit is not permissible."

Garevski lays the blame on a number of factors. Certainly, locations were wrong. Some buildings were placed over an active fault, some on soil susceptible to liquefaction (so subsidence overturned them) and some were built on unstable coastal sites and disappeared into the sea along with the land. In those circumstances, he says, even the strictest application of seismic regulations and construction codes will not prevent failure. Better land-use guidelines are needed to ensure buildings are put up on solid ground.

But bad locations were not the main cause of the damage in Turkey. The primary reasons, according to Garevski, were a structural system which should not be applied in a high-seismicity region; bad quality concrete; and improper reinforcement. Istanbul does at least stand on solid ground. Polat Gülkan, of the Middle East Technical University in Ankara, points out that much of the city is built on "competent soil" – soil that will not liquefy. But, as in the Marmara region, the building materials typical to Istanbul are reinforced concrete frames with masonry infill. The reinforcement has not always been designed to withstand the special stresses of an earthquake, and both concrete and masonry have all too often been substandard. Newer steel-framed high-rises and, paradoxically, jerry-built shantytown dwellings, are considered less vulnerable.

Turkey is among a dozen countries included in a 2001 risk analysis for the Stability Pact for South Eastern Europe. Its Disaster Preparedness and Prevention Initiative (DPPI), a Stability Pact effort to help develop a cohesive regional strategy, underlines "the significant role that proper construction can play" in reducing quake damage. But its operational team – drawn from the United Nations Development Programme (UNDP), the

North Atlantic Treaty Organization (NATO), the United States Agency for International Development (USAID), the International Federation, the Swedish Rescue Services Agency, and Italian, Bulgarian and Croatian governmental bodies – concludes that enforcement programmes have been neglected or abandoned, resulting in another generation of housing and industrial facilities being built without minimum protection.

In south-eastern Europe, part of this may relate to the transition from communism, from command to free-market economies in some states. The change does bring some benefits. Disaster preparedness is no longer the exclusive preserve of the military, and better trained, better equipped civilian units are now involved. But economic reform and structural adjustment have stretched economies and capacities, and will continue to for some time. Health and social welfare have suffered as a consequence, so it's hardly surprising that the enforcement of building regulations – not the easiest operation at the best of times – should also feel the pinch. Observers say widespread bribery and corruption throughout the region exacerbate the situation.

Macedonia is a country that may be in the process of coming full circle. While strict enforcement of regulations existed under socialism, particularly after the horrific 1963 Skopje earthquake, the dissolution of the former Yugoslavia allowed rules to be ignored. "There was no one to ensure there were controls," says Sune Follin, the International Federation's regional disaster preparedness delegate and DPPI team leader. "There was no follow up. Nothing was happening. House building today may be no problem, but in the early 1990s who knows? Construction was worse then than anything that had come before."

## Initiatives needed to enforce building codes

Turkey's building regulations are fine. Its construction code has been on the statute books since the catastrophic Erzincan earthquake of 1939 and revised several times, the last in 1997. It prescribes an impressive chain of inspections from the planning phase onwards. Unfortunately these are often quite simply ignored. There are too few trained inspectors, and shoddy building practices and corner-cutting remain common. Some lessons have been taken to heart. In the Marmara region, the International Federation has constructed or renovated five schools and five hospitals to earthquake-resistant standards. But how do you ensure such practice becomes commonplace?

Improving building practice will require both sanctions and encouragement. Here, says Follin, there is an important role for the insurance industry, working with national authorities to provide incentives for adherence to building codes – some carrots to accompany the legal sticks. They could also share some of the risks. New Zealand, for example, has used both insurance and tax policies to encourage better building practices. In Turkey neither has played an active role. Swedish development analyst Ian

## Box 5.2 Iceland – model of efficient disaster preparedness

*What you expect, you can prepare for.*
*Icelandic saga*

At the north-western corner of Europe, far out in the Atlantic Ocean, directly on the fault line where the American and European continents meet, lies the volcanic island of Iceland, one of the most disaster-ridden places on earth. Tremors are almost daily events, and few years go by without a sizable earthquake. Volcanic activity is almost constant: the last major explosion was just offshore in 1973. Vulcanologists expect another almost any day. To complete the picture, Iceland is also subject to avalanches and extremely violent storms.

This disaster-prone country has, however, learned to deal with its natural hazards remarkably well. Iceland lacks an army, but has a well-integrated system of response, comprising the civil defence, the Icelandic Red Cross, police, firefighters, the independent search-and-rescue association and special auxiliaries including the scouts. All across the island, even in sparsely populated areas, disaster relief centres have been designated with trained leaders, often teachers. The goal is to have at least eight people fully trained for emergency response in each centre. Both the Red Cross (represented in 51 districts) and the search-and-rescue (S&R) association are entirely made up of volunteers. S&R members are trained in the use of heavy equipment. Psychological preparedness has become a priority for the Red Cross, which is teaching its volunteers informal counselling techniques.

Teachers, many of whom are involved in disaster preparedness programmes, are at the heart of effective disaster education. Even nursery-school children are taught how to behave in an earthquake: for example, to seek shelter under the nearest solid table. It becomes second nature. Can other countries emulate what Iceland has done? There are cultural factors at work here, which are not necessarily present in other earthquake zones. Iceland is a rich and small society. Like the other Nordic countries, Iceland enjoys almost 100 per cent literacy, which makes education programmes easier. It is also a disciplined society. It has been able to institute a strict building code which is largely adhered to. And a culture of civic responsibility underpins its extensive network of volunteers. ■

Christoplos points out that "an ambiguity emerges when one shifts from total control to a free market. Officials and politicians are unsure where they should intervene in market mechanisms, for example: how to mesh regulation with commercial pressures from insurers".

Before 1999, the Turkish government paid, more or less, for all earthquake damage, and until a few years ago private earthquake insurance was forbidden. After the very costly 1992 Adana quakes, the World Bank persuaded the government to reverse its policy and insurance became mandatory for housing. The law, however, is not applied until a property is sold, when insurance becomes obligatory for the new owner. Moreover, premiums are calculated purely on location, by earthquake zone, and are not differentiated to provide incentives for better building practices. The government did decide to hold contractors responsible for the quality of new buildings and imposed a ten-year liability. Unfortunately that decision was reversed by the High Court and never became law.

The insurance industry would seem to be relieved. The deputy director of the Turkish catastrophe insurance pool, Barboros Yalcin, is candid: "As compulsory earthquake insurance managers," he says, "we don't want to write many more policies in Istanbul" – where 40 per cent of all Turkish policies are already based.

Perhaps the greatest inducement to good practice is better understanding among those who actually do the designing and building. With leading research centres like Turkey's Middle East Technical University, Macedonia's IZIIS and the National Building Research Institute in Romania, there is no shortage of knowledge in south-eastern Europe. But the knowledge is insufficiently shared.

IZIIS, a recognized international training centre, says earthquake engineering has not been adequately included in any faculty programme anywhere in the region. Semih Tezcan, professor of civil engineering at Bosphorus University, deplores the fact that many architects and civil engineers can practise without having had a single earthquake-related course in their university training. He would like to see the national curriculum modified to change that.

## Strengthen "lifeline" infrastructure

For buildings already at risk, the best mitigation option is known as "seismic repair" or "retrofitting" – reinforcing the structure to make it more earthquake-resistant. The big question here is economic, and many governments are either unable or unwilling to foot the bill. Some argue that it isn't the task of government to pay for retrofitting, but to enforce regulations so that those who own buildings are forced to foot the bill. Either way, retrofitting is expensive and a cost/benefit calculation is essential.

In Istanbul, it's highly unlikely that all its buildings could be made earthquake-resistant, although that does not mean there is nothing to be done. One mitigation option is to identify, assess and reinforce "lifeline" infrastructure essential for public safety, such as schools, hospitals and indoor sports centres – the latter not for sports but to serve as better public shelter than tents, especially in a cold and rainy Turkish winter. Tezcan has calculated that assessment would cost US$ 3 per square metre. Retrofitting costs more, but assessment should help to decide what is worth working on and what is not.

The need to regulate the retrofitting of existing buildings is well illustrated by assessments in Turkey and Greece. These show that an enormous number of structures,

chapter 5

which appear to have come through tremors unscathed, are nonetheless below legal standards. In some cases they may have been weakened and left more vulnerable to future quakes. Macedonia is another example: a great number of its old buildings risk severe damage in future earthquakes. Prompted by these concerns, a large joint seismic assessment and rehabilitation project is now getting under way in all three countries, financed by NATO. The knowledge acquired should help establish regulations for retrofitting throughout the region.

Retrofitting private dwellings, however, can prove very complicated. A large proportion of flats in Istanbul, for example, are condominiums. For any major work of strengthening, or even repair, all the owners must agree to it. Even if they do, it is often difficult to find either financing or alternative housing while the work is being done. Landlords are no keener.

Polat Gülkan argues that: "Careful policies for encouraging building upgrades, bundled with tax breaks, cheap loans and other incentives are needed, even for demonstration-size programmes." But, he adds, "an average building in Turkey has a lifespan of 50 years, [so] there is a strong likelihood that many may reach the end of their lifetimes without being tested." The most realistic way forward, suggests Gülkan, would be to replace substandard housing once its design lifespan has ended, as long as building codes are adhered to.

Of the 50,000 people pulled from the rubble of Turkey's Marmara quakes, 98 per cent were rescued by locals. Investing in municipal-level disaster preparedness will save more lives.

Mikkel Ostergaard/
International
Federation, Turkey
1999.

## Box 5.3 First aid: humanity in action

In a local school in Posoltega, Nicaragua, children remember with fear the morning when a landslide from the nearby volcano consumed much of their village. Some 1,400 people lost their lives in that one disaster, triggered by Hurricane Mitch's torrential rains. Since then, the Red Cross has been targeting schools with first-aid training. After just two-days' training for six teachers and 70 children, Posoltega's school was ready to carry out an earthquake response simulation exercise. The older children gave first-aid care to 12 children with differing injuries and then evacuated them. At the end of the event, the children felt proud of their skills and better prepared to deal with future health- or life-threatening situations.

Posoltega's landslide was exceptionally devastating. But first aid is relevant not only for the big, once-in-a-lifetime disasters – it is equally important in dealing with everyday problems and hazards. Dolisie is a city of 70,000 in the southern part of conflict-torn People's Republic of Congo. Retraining of Red Cross first aiders formed a key part of the relief and reconstruction effort. These volunteers have since been involved in disease prevention, sanitation and hygiene campaigns. Recently they were a valuable trained resource in assisting the many victims of a serious railway crash. Equally, Red Cross Red Crescent teams are mobilized to provide first aid during pilgrimages (e.g., Mecca, Saudi Arabia), sports events (such as the football World Cup in France), festivals (e.g., summer of music, Argentina), street demonstrations (e.g., Philippines) and so on. Teams are also mobilized daily to provide first-aid posts for small community events and for training activities.

Whether during a major public disaster or a small private event, injured people often feel too overwhelmed to cope. There may be a time lag between the injury and the arrival of professional help – a period during which people on the spot have to take action alone. Whether they are community members or Red Cross Red Crescent staff and volunteers, they need to be prepared for the worst. So it is essential to develop the individual capacity of every person to help protect and save lives through first aid. According to the Armenian Red Cross's first-aid national coordinator, "These actions of normal people in exceptional circumstances are possible because they get the confidence and skills to act on a daily basis through first-aid training and services."

The efforts of Red Cross and Red Crescent societies to promote and disseminate first aid represents one of their major activities. Their first-aid programmes are tailored to account for communities' specific needs and capacities. Results are even better when first-aid training includes preventive measures. Training and mobilizing communities and volunteers through first-aid activities has two outcomes. Firstly, by reducing the impact of daily crises, first aid can help promote the sustainable development of the community. Secondly, it can prevent and alleviate the suffering caused by more high-visibility disasters. First aid is humanity in action. And it is disaster preparedness in its most versatile and fundamental form. ∎

# Preparing for earthquakes

Mitigating earthquake risk by enforcing building codes and retrofitting lifeline infrastructure will take time, money and political will. It is a long-term strategy. But what if catastrophe strikes before enforcement takes effect, or before the 50-year design-life of poorly constructed buildings expires? National governments, municipal authorities and exposed communities can all take action now to prepare for the worst. Lives can be saved if all those at risk understand the quickest and most effective ways to respond to disaster (see Box 5.1).

There are promising signs of progress in disaster preparedness. Since Turkey's devastating Marmara quakes in 1999, new crisis management centres have been established in Istanbul and Ankara. Both are impressive installations, fully equipped and ready to deal with the next large-scale disasters whatever they may be. In the case of a serious Istanbul earthquake, a direct electromagnetic connection would shut down gas and electricity distribution instantly.

Then, so the plan goes, high-level representatives of all the services involved would gather at the centre: military, municipality, departments of public works and health, administrations of 32 subdistricts, and the Turkish Red Crescent. Army helicopters would take off to assess damage and start ferrying the wounded to hospital.

Istanbul has been carefully mapped by geographic information systems and the detail, street by street, practically house by house, is impressive – alternative routes for emergency vehicles, space for a million tents, even emergency graveyards. The assumption is that the two bridges across the Bosphorus would hold. These are vital, but in case there are problems with access roads, the government could commandeer all the boats in harbours and marinas, including 2,000 yachts.

How well all these excellent plans can function, if necessary, is not known at the moment. A recent simulation exercise scared everyone involved. The death and injury figures were horrific, and those responsible have returned to their drawing boards to do some revision.

## Define and decentralize disaster roles

A key priority in preparing for disaster is planning. That means deciding on the most effective strategy to pursue when disaster hits. Within that strategy, the roles and responsibilities of different agencies must be defined. Otherwise, the chaos of disaster will create a chaotic response.

Experience from recent European earthquakes testifies to this. The free-for-all nightmare was abundantly clear in the 1999 Turkish quakes, nowhere more so than in the small town of Kaynasli, 100 kilometres east of Istanbul, the epicentre of the second one. Two years on, some of its citizens undertook an evaluation project and their report, *We survived the earthquake,* is a recapitulation of what has been learned, much of it applicable to larger cities.

Some of its most original thinking concerns the role of non-governmental organizations (NGOs), whose numbers and range of concerns have shot up in Turkey in recent years. The *We survived* group has thought hard, and with some bitterness, about the 2,000 NGOs which poured in after the earthquake. Some, of course, were useful. But others are described as "charlatans". The report suggests the establishment of a register of genuine NGOs. According to Turkish law on disasters, NGOs can claim reimbursement from the government for all relief expenditure, explaining why much more money is spent on relief than on preparedness. The Kaynasli group suspects it is also an unfortunate motivation for some NGOs who "come sweeping in like vultures".

The question of whose responsibility it is to prepare for and respond to disasters isn't always easily answered. There are governments that preclude a proper delegation of duties. The Stability Pact's DPPI risk analysis finds that in south-eastern Europe, the operational plans and management structure that accompany them are often too centralized and difficult to implement on a local basis at the disaster site. "Highly centralized systems of governmental authority and allocation of resources often create delays and add layers of bureaucracy, compounding problems of an already difficult emergency response situation," it comments. Countries should critically review if and how decentralization would benefit their operations.

Disturbingly, the DPPI analysis adds that, although most of the states which its teams visited possess some form of national disaster management plan, few appear to be comprehensive.

"They do not define clear roles for individual organizations," the report says. Nor do they provide "an adequate base for mutual support from others within the nation or external support from neighbouring nations". According to the analysis, "In almost every country visited, the Red Cross or the Red Crescent society proved to be a strong, well-organized capability that was known and trusted by both the public and the government. In several cases they provided the only real pre-disaster planning, organization and response resource."

National Red Cross and Red Crescent Societies, however, should act as *auxiliaries* to government rather than instead of government. "It is imperative that the Red Cross or Red Crescent role in a national disaster plan is clearly defined, as the roles of government agencies and NGOs should be," maintains Sune Follin. "Where this has not been the case, there has been calamity." If the situation demands it, National Societies can in turn fall back on regional and international Red Cross Red Crescent capacities. But training teams at municipal and local levels to respond to disaster is the most effective way to save lives.

## Local emergency response saves lives

Whenever there is a serious earthquake, especially in a city, international teams fly to the rescue, often with heavy, expensive equipment and teams of search dogs. Serious controversy now surrounds these high-profile interventions. Does it make sense for outsiders to conduct search and rescue after earthquakes? Oktay Erguner, director of Turkey's national crisis management centre, points out that it is rare for international teams to arrive on the site of an earthquake quickly enough to be really effective. After a couple of days, survivors are found only in exceptional cases.

Of course, no effort must be spared to save any life. But disaster experts are questioning the sense of sending teams to locations when a critical time lapse is inevitable. It makes good television, but wouldn't the money and effort be better spent training local people in earthquake zones in simple emergency response? According to Erguner, at least 50,000 people were found alive under collapsed buildings after the Marmara earthquakes. Neighbours and local people rescued 98 per cent of these. Outside professionals rescued just 350. So decentralizing disaster preparedness and response resources to municipal and local levels will bring the greatest benefits (see Box 5.2).

However, local emergency response often depends on volunteers – who must maintain their skills and commitment in between disasters. Bureaucracy and boredom both conspire to undermine their effectiveness. The key question facing the Red Cross Red Crescent, and other organizations dependent on volunteers, is not merely how do you prepare for the inevitable, but how do you keep your people motivated? "A major challenge is keeping them warm," says Sune Follin. "It's all very well preparing. It is easy to raise initial interest, particularly like now in Romania or Albania where people are well aware a catastrophe could be around the next corner. But where there can be quarter or half a century between major earthquakes, interest can wane. Half the people you train may be dead of natural causes by the time the big one hits."

Follin questions whether training Red Cross Red Crescent volunteers in specialist urban activities such as search and rescue is wise, if the professional skills and alertness which must be maintained are rarely called upon. Nor, he argues, is it making the most of crucial human resources. "Training people to be multi-purpose, as indispensable in dealing with traffic accidents as they are in mountain rescue or earthquakes, makes far greater sense," he maintains.

## Box 5.4 Countering the risks of a Kathmandu quake

The scenario is frightening. In 1934, over 16,000 people died in Nepal's Kathmandu valley during the Great Bihar earthquake, which measured 8.4 on the Richter scale. If a similar quake struck today, it would leave around 40,000 people dead and another 95,000 injured. Between 600,000-900,000 would be rendered homeless. More than 60 per cent of the valley's buildings would be destroyed.

Estimates suggest that 95 per cent of water pipes, and half the pumping stations and treatment plants would be seriously affected for several months. All electricity stations, nearly 40 per cent of electricity lines and around 60 per cent of telephone lines would be out of order for up to a month. Roads and bridges would be severely damaged, isolating Kathmandu's international airport. Consequently the arrival of international relief assistance would be seriously hampered.

"This is not merely sensationalism. The devastating impact of the recent Gujarat earthquake in India has confirmed that these estimates are not merely academic," warns Abod Mani Dixit, general secretary of the National Society for Earthquake Technology-Nepal (NSET). For Dixit, a geologist by training, the quest to reduce seismic risks began when a schoolchild asked him about the consequences of a quake for the Kathmandu valley. That was in 1992. Since then, NSET has brought numerous international agencies, the Nepal government and civil society organizations together to create a long-term disaster action plan to reduce earthquake risks in the Kathmandu valley. The plan was formulated through a series of workshops, conferences and, unusually, regular public hearings.

Nepal is situated in the seismically active Himalayan mountain belt. More than 1,000 tremors ranging from 2 to 5 on the Richter scale rock the mountain kingdom every year. Worse still, the Kathmandu valley is actually a lake basin consisting of soft sediments susceptible to liquefaction. And the valley's "basin effect" amplifies seismic waves, increasing their destructive potential. Studies conducted by the UN for 21 highly vulnerable cities around the world list Kathmandu as the city at greatest risk.

Dixit's risk reduction strategy combines both disaster mitigation measures (e.g., strengthening buildings) and preparedness measures (such as disaster awareness and training). "We are following a two-pronged approach, preparing to lessen the impact of the impending disaster and to prepare the community's capacity to cope and fight in the aftermath of the disaster," he says.

After a series of workshops, NSET began assessing schools in the valley to determine the extent of vulnerability, the techniques to be used in strengthening buildings and the costs involved. Based on this assessment, NSET has to date structurally reinforced nearly a dozen school buildings to make them earthquake-resistant. To enhance disaster preparedness, NSET has ensured that each school prepares an emergency plan with duties and responsibilities assigned to staff members. Earthquake risk reduction concepts have been woven into the school curriculum. Regular "duck and cover" and emergency drills are conducted with students and teachers, periodically monitored by the relevant authorities. Through schoolchildren, the principles of earthquake preparedness also reach parents.

Hospitals face enormous pressure from the large numbers likely to be injured during a seismic disaster. Yet many of Kathmandu's hospital buildings could themselves collapse in the event of a major earthquake. NSET, in collaboration with the World Health Organization (WHO), has extended its programme to improve the seismic resistance of hospital buildings. On the basis of NSET's assessment, a specialist engineer from Ecuador was invited to visit 15 hospitals and recommend structural improvements. WHO will contribute towards the cost of the changes. Meanwhile, 20 one-day seminars are planned to train 400 key staff members from major hospital emergency wards in disaster logistics, triage technique and emergency medicine.

Dixit says that the school and hospital projects have created a much wider awareness of the issues related to earthquake risk in the valley. Kathmandu residents are now seeking NSET's guidance in reconstructing their homes to withstand seismic shocks. There has been growing interest from the government, universities and civil society to improve the risk management capacities of public utilities and emergency services. Since the Gujarat earthquake, NSET has organized 50 training sessions in disaster preparedness for the police, army, embassies and a score of other crucial agencies. And the Nepal Red Cross Society is planning to bring its emergency response expertise to the initiative.

To help people understand the threat they are faced with, NSET disseminates information through a series of simple, illustrated publications which detail likely scenarios one day, one week, one month and one year after disaster strikes. Such "software" measures have greatly helped spread the risk reduction message throughout the valley. The exemplary work undertaken by Dixit and his team has been used as a model by the UN in nine other cities around the world.

Meanwhile, the UN's own disaster management team (UNDMT) is actively developing its readiness to confront a major earthquake. Emergency planning is critical given the fact that Kathmandu is the main centre for all the health and utility services in the area. If the city takes a direct hit, the possibility of immediate relief coming from remote mountain villages on the valley rim is very remote. So the UN's contingency plans account for providing, within 24 hours, initial survival assistance in the form of food aid, health care, water and shelter for 200,000 people for a week. This could extend to providing relief for up to 500,000 people for a month.

The message that disaster preparedness is the key to survival in any eventuality is engraved on a unique monument to the 1934 earthquake that stands in the heart of the Nepalese capital. Kathmandu is perhaps the only city in the world with a statue that forewarns society to be prepared for natural disasters. And rightly so. ■

His argument is backed up by the fact that, as needs have increased around the world, organizations dependent on voluntary work have seen volunteer numbers decrease. It isn't enough to train people. They must be well managed and they must put their training to good use in order to stay motivated. Without motivation, volunteers will melt away.

In Slovenia's national protection and rescue system, the Red Cross has been tasked with providing basic assistance to help the population survive a catastrophe. Red Cross emergency response units have been established in Ljubljana, the capital, and two other towns, ready for deployment within a few hours of a natural or man-made disaster. The units will back up the medical services and provide food, drinking water, care and shelter for the most vulnerable, such as women with children, pregnant women, the disabled and the elderly.

Slovenia's emergency response units are also prepared for an earthquake. Last year in an exercise, civil defence chiefs requested their immediate deployment after an imaginary quake left 474 people dead and more than 5,000 injured near Ljubljana. But for the volunteers, earthquakes are just one of a range of disasters in which they could intervene. Others include floods, landslides, dam bursts, unexploded ordnance, transport of hazardous material, industrial and nuclear accidents. There is no waiting for the big one here; volunteers' multi-purpose skills are constantly in demand.

Like the Slovenes, the Romanian Red Cross prioritizes preparedness for a range of disasters. It has trained more than 4,000 volunteers for 278 intervention teams which can be called upon at any time for any kind of disaster. Floods are an annual occurrence, as in 2000 when the Tisza River and its tributaries burst their banks in the north-west, overwhelming close to 15,000 people in one of the poorest corners of the country. The Red Cross emerged as the primary partner of government in delivering assistance.

Perhaps the most versatile of disaster preparedness skills is first-aid training, in which Red Cross Red Crescent societies specialize. Volunteers put their first-aid training to use all over the world, not only during disasters but at a range of diverse events (see Box 5.3).

## Converting ideas into action

Keeping volunteers warm is one thing. Keeping politicians and governments warm is another. "We are the most popular people in the world after an earthquake," says a leading European seismologist. "Politicians are all over us. They ask how they can ensure such a tragedy never happens again. What will it cost? What can they do for us? But then it tails off, and that is one of our major problems. You need an earthquake every five years to keep politicians interested."

The recipe for reducing risks from earthquakes, and from a range of other natural disasters, is nothing new. The Stability Pact's DPPI assessment places building codes atop its list of factors considered vital for the reduction of seismic vulnerability. Other factors include public awareness; enforcement of appropriate land-use guidelines; population distribution; public infrastructure; and effective warning, evacuation and

chapter 5

response procedures. "In most cases," it reports, "nations possess considerable scientific capability to identify, assess and delineate high-risk areas, and design appropriate national codes and regulations. What they lack are the administrative instruments to implement and enforce effective programmes exploiting this scientific insight."

So, are national governments shirking their responsibilities to ensure safer environments for their people? Are there more proactive ways of converting the lessons of the past into the achievements of the present – *before* the next disaster strikes?

How countries respond to earthquakes, of course, only reflects their approach to wider risk reduction, and in south-eastern Europe, says the assessment, there is a considerable range. "Some are quite sophisticated and have well-developed systems of preparedness and prevention that seem pragmatic and sustainable. They appear to anticipate problems and reflect good risk assessment and planning methods. Others appear to be in a purely reactive mode waiting for something to happen, hedging on necessary political or financial investments, simply playing the percentage game and hoping that disasters might not occur."

The tide, however, would appear to be turning, and the DPPI assessment detects growing recognition that disaster prevention and emergency response must be a priority function of governments. But there is a long way to go. Sune Follin refers to a disconcerting trend in some countries that simply passing national disaster laws is considered a sufficient response. "What is missing all too frequently is enforcement, or the resources it takes," says Follin. But there are others like Albania where the government is more forceful, he says: "People there understand that laws are meaningless until they are backed up by aggressive enforcement."

Albanian legislation introduced in 1998 has tightened high-rise construction and any building of more than eight storeys requires the approval of the national seismological institute. The modification of ground-floor dwellings in apartment buildings for commercial purposes must have the approval of a government-mandated engineer to ensure reinforcement is unaffected. In the past, structures have allegedly been weakened by, for example, the need to create space in shops or restaurants. Without the approval, municipalities will not consider an application. Other legislation has sought to sharpen inspection.

## Public awareness promotes culture of prevention

Political imperatives to act can be created, and public opinion and media coverage are powerful incentives. During a European Community (EC) workshop on earthquake risk in 2000, Commissioner Wallstrom argued that "as news arrives almost instantaneously at everyone's home, public opinion is requesting a better and safer European environment". The workshop cited "lack of public awareness" as a key factor in increasing the vulnerability of modern cities to earthquake disasters. Humanitarian organizations and NGOS can play a key role in promoting a culture of risk reduction and prevention through advocacy campaigns targeted at both decision-makers and exposed communities (see Box 5.4).

In central and south-eastern Europe, the International Federation is urging National Red Cross and Red Crescent Societies to see the media as partners in disaster mitigation and preparedness. Their role could be to contribute towards public awareness of dangers and how to respond to them. Or it could be to advocate for changes in government policy. This means fostering strategic links between humanitarian organizations and local and international media, the development of dialogue with publishers, editors, programme makers and correspondents, and the co-opting of local journalists into working groups on risk reduction.

"Working with the media isn't only about ensuring profile for your organization. It is part of how we can change things for the better," argues Follin. "What the media wants is a story. How the community can ensure 1,650 people need not die, and 10,000 people need not be injured the next time an earthquake hits Bucharest, is a great story. The dialogue we seek is a little more nuanced than that, of course. It is about us providing insight and opportunity, and the media fulfilling its responsibility to the community. The bottom line is public awareness and momentum."

More than two years after the Marmara earthquakes in Turkey, hardly a day goes by without the Turkish media carrying a story related to earthquakes. Whipping up hysteria has become a favourite pastime of some parts of the popular press. Others, such as Istanbul's private Open Radio – favoured by students and the intellectual elite – shoulder responsibility and regularly produce thoughtful programmes devoted to the subject, with invited experts and air time for the public to express their concerns. They realize how thin the line is between building public awareness and preparedness and feeding anxiety by, for example, focusing on how many houses would collapse. In such quarters, strong allies can be found. Others need converting.

## Regional cooperation raises standards

Partnerships to reduce disaster risk can transcend borders. The severity of earthquakes, the likelihood that the coping capacity of any one nation will be overwhelmed, the sudden and urgent need for external assistance: these are powerful arguments for neighbours to help one another.

Even politics and traditional tensions can make way for cooperation in time of disaster. Following the Marmara earthquakes, Greece, with whom the Turks have sometimes difficult relations, offered immediate assistance. This year, the Greek foreign minister George Papandreou once again responded. "We express our solidarity toward the troubled people and government of Turkey who have again been struck hard by another earthquake," he said.

Such developments interest the Stability Pact, and on more than one level. Where nations share risks and vulnerabilities, the foundations for sustained cooperation in disaster mitigation and preparedness can be found. The identification of such risks was the rationale for the DPPI regional assessment. How does a nation weigh the risk of a disaster that may occur only once or twice a century against annual or frequent catastrophes? Where should it place emphasis? Seen against the full range of natural disaster risks faced by a region, options become clearer.

South-eastern Europe has most of the hazards imaginable – floods, forest fires, earthquakes, recurrent landslides, droughts, storms and a number of man-made threats. The assessment has sought not only to establish the common risks, but to determine how vulnerable to them each nation is, the level of public awareness, the availability of means to communicate the risk, and the political and administrative environment in which risks are dealt with. Insight into these, it says, will help it decide where to concentrate efforts, and direct maximum resources to improving mitigation, preparedness and response. The perspective will also better define how nations can assist each other.

The DPPI team is encouraged by what it describes as "opportunities for more adequate multilateral coordination" and mutual assistance between neighbours. While every country has weaknesses, they also have strengths that are the basis for regional cooperation. National disaster management plans, the assessment concludes, should ultimately reflect an agreed-upon regional standard that would facilitate collaboration.

It is hard to think of a region where this philosophy makes greater sense than the highly seismic Balkans. IZIIS estimates that average losses due to recent earthquakes there amount to 20-30 million euros (US\$ 17.5-26.5 million) per year, apart from the human casualties. Therefore development of a seismic risk reduction strategy has to be a regional long-term objective.

## Active humanitarian advocacy

Regional cooperation in disaster mitigation, preparedness and response, knowledge sharing, and the introduction and enforcement of regional standards, particularly in construction, clearly hold the promise of dividends. Lessons painfully learned should not be forgotten.

Expert reports of the Izmit disaster refer, among other things, to the width of traffic arteries. Some collapsed buildings blocked the way for rescuers through narrow streets. Others toppled buildings on the opposite side of the street in a domino effect. Macedonia's Professor Garevski emphasizes that free access for rescue services is vital in emergency situations – a lesson of particular importance for the designers of future urban plans in seismic regions.

The lesson wasn't new. Getting on for four decades previously, Skopje, where Garevski's own institute is situated, learnt the hard way. Since that catastrophe in 1963, the rebuilding of the Macedonian capital has left it much better off in terms of mitigation. Enlightened planning and legislation produced wider streets, lower population densities and more effective building regulations.

It doesn't take high-tech mapping to remember recent lessons. Common sense derived from experience is sufficient, plus the political will to act upon it. But do cities first need a disaster before acting? Do humanitarian organizations wait to soothe the suffering in the aftermath of disaster? Or should they seek to reduce the risk of disaster striking in the first place?

*Strategy 2010,* the Red Cross and Red Crescent's plan for the decade, calls for more active humanitarian advocacy. In the context of urban earthquake risk, that means helping to create ever-greater public awareness of the threat, more dialogue with authorities on how to mitigate the threat, and better preparedness in the event of the threat becoming reality. Advocacy not only aims to create awareness; it aims to achieve real changes both in people's behaviour and in government policy. Critical changes that will reduce the risk of future earthquakes include:

- **Legislation and enforcement** of regional standards in construction, land-use and urban planning – including incentives to encourage better building.
- **Decentralized disaster preparedness** and response planning, along with the resources to train emergency teams in exposed communities.
- **Regional knowledge sharing**, to make sure what is learnt in one place is known in another, through specialist training and public information.
- **Promoting responsible public debate** through national media to maintain the pressure on policy-makers to prioritize risk reduction.
- **Raising public awareness** of the threats and how to react to them, through mass media and education of schoolchildren.

*Principal contributors to this chapter were John Sparrow, head of regional communications at the International Federation's Central Europe delegation in Budapest, and Liesl Graz, an independent writer based in Switzerland. Boxes 5.1 and 5.2 were contributed by Liesl Graz; Eric Bernes and Stephen Claffey, both with the International Federation, contributed Box 5.3; Devinder Sharma, a journalist based in India, contributed Box 5.4.*

## Sources and further information

Bendimerad, Fouad. *Building Disaster Resilient Megacities.* 2000.

Bishop, Joseph A. *Rapid Study Report on the International Search and Rescue Response to the Izmit Earthquake.* Gibraltar, 1999.

Special Co-ordinator of the Stability Pact for South Eastern Europe, Disaster Preparedness and Prevention Initiative. *Regional Report of the DPPI Operational Team: The Gorizia Document.* May 2001.

Garevski, Prof. Dr. Mihail. *Damage to building stock due to the 1999 Izmit earthquake.* 2001.

## Web sites

Hellenic Red Cross **http://www.redcross.gr/**

Icelandic Red Cross **http://www.redcross.is/**

Institute of Earthquake Engineering and Engineering Seismology (IZIIS) **http://www.iziis.ukim.edu.mk/**

Middle East Technical University **http://www.metu.edu.tr/**

National Society for Earthquake Technology-Nepal (NSET) **http://www.geohaz.org/project/kv/nsetdesc.htm**

Stability Pact for South Eastern Europe **http://www.stabilitypact.org/**

Turkish Red Crescent **http://www.kizilay.org.tr/**

Section Two

**Tracking
the system**

**chapter 6**

# Assessing vulnerabilities and capacities – during peace and war

Designing a disaster preparedness plan is impossible without knowing the risks facing any given community. But unlike hazard mapping, which can be computerized using geographic information systems or even satellite images, vulnerability cannot be seen from above. Vulnerability changes constantly, reflecting prevailing social, economic, cultural and political circumstances. And yet, it is this same vulnerability that can dramatically intensify the effects of a disaster. Similarly, the capacities of communities to cope with hazards and disasters will vary according to local conditions and perceptions.

Various organizations specializing in development came to recognize the importance of understanding capacities as well as vulnerabilities in the late 1970s. This recognition is now increasingly mirrored in humanitarian circles, aware of the limitations of needs-based approaches, which often focus on material aid and regard beneficiaries as passive "victims". In 1989, Mary B. Anderson and Peter J. Woodrow presented their guidelines for analysing capacities and vulnerabilities, known as CVA (see Box 6.1). And in 1994, the *Code of Conduct for the International Red Cross and Red Crescent Movement and Non-Governmental Organizations in Disaster Response* emphasized the need to build disaster response on local capacities. These initiatives formed part of the conceptual basis for the International Federation's approach to disaster risk planning, known as the vulnerability and capacity assessment (VCA). Its main aims are threefold:

- To assess the risks facing communities and the capacities available to deal with those risks.
- To involve communities, local authorities and humanitarian/development organizations in the assessment from the very outset.
- To draw up an action plan to prepare for and respond to the risks identified.

Involving all parties from the outset is fundamentally important, for two reasons. When risk reduction plans and projects are finally introduced on the recommendations of the VCA, the partnership process which the VCA generated will of itself improve cooperation during the implementation of these projects. Perhaps more importantly, the assessment can provide participants with greater awareness of their own potentialities. "Instead of seeing themselves as victims, people tell themselves that they can influence what happens," says Graham Betts-Symonds, senior officer for disaster preparedness at the International Federation. So VCA is not just a diagnostic measure, but a capacity-building tool, because it involves communities from the very start.

Photo opposite page: Breaking new ground: the Palestine Red Crescent's vulnerability and capacity assessment took into account what children view as hazards.

© Howard Davies/ Exile Images, West Bank, 1992.

130

Crucially, the VCA process aims to heighten the awareness of communities and aid organizations *before* disaster strikes. It then converts that awareness into concrete, pre-emptive action: mapping hazards, vulnerabilities and capacities; agreeing on the roles of organizations and community members during disasters; and training in disaster response and first aid. In this way, VCA plays a key role in triggering and supporting disaster preparedness and mitigation initiatives.

However, carrying out a detailed investigation of the risks facing a community or even a nation, and the abilities of people and organizations to meet those risks, can be a daunting, controversial and delicate operation. The assessment may expose risks rooted in long-established inequalities of access to resources and power. The VCA's conclusions could be deeply political and unacceptable to national authorities, or even to the leadership of the organization conducting the VCA. The assessment process risks raising unrealistic expectations among vulnerable people that their problems may vanish. It may end up creating a wish-list of priorities beyond the capacities of national organizations to deliver.

## Early experiences

To date, hundreds of VCAs have been completed at community level. Some 30 National Red Cross and Red Crescent Societies from Scandinavia to Africa have embarked on national-level VCAs and 15 more have declared their intention to start doing so during 2002 (see Box 6.2). The Swedish Red Cross (SRC) was one of the first to experiment. They had two aims: better identification of vulnerable groups; and greater awareness and ownership of the VCA process itself on the part of SRC's staff and volunteers. This double-edged aspect of VCA – understanding "clients" better *and* improving the skills of the National Society carrying out the assessment – is still central to the assessment's approach.

The VCA's popularity may be because its essence is close to the community-based way in which National Societies operate. A recent article by Christoplos, Mitchell and Liljelund expresses this as follows: "Organisations such as Red Cross/Crescent Societies, religious institutions and farmer organisations have a presence at grass-roots level and are ideally placed to develop an understanding and knowledge of local capacities and coping strategies and, moreover, to involve communities in shaping disaster mitigation and preparedness activities." This is particularly relevant for communities isolated by geography or political realities from the support of their national governments (see Box 6.3).

A good VCA can be long, exhausting, and expensive. It requires total commitment on the part of both collaborators and volunteers (see Box 6.4). But the effort pays off: witness the experience of The Gambia Red Cross (GRC), whose 18-month-long VCA was completed in 1998 and had several significant outcomes:

- It helped the GRC identify 22 potential hazards (they were expecting only about half that many), which allowed them to develop a strategic workplan for 2001-03. Surprises included seasonal food shortages, waste dumping, meningitis and drought.
- Two of the hazards identified (HIV/AIDS and traffic accidents) are now the focus of major projects.
- It helped the GRC discover which communities had stronger capacities than others to deal with disasters on their own.
- The GRC's work has become more focused and the Red Cross's presence and profile in the community has grown.
- The GRC is now a member of a working group, based on the VCA's results and set up by the government, which is compiling a national disaster management and contingency plan.

## VCA under fire

Conflict situations make carrying out a VCA particularly challenging. The experience of the Palestine Red Crescent Society (PRCS) with VCA has shown, however, that the process can make a difference whether during peace or war.

On 28 September 2000, the Palestinian territories erupted, marking the beginning of what became known as the second "intifada" (uprising). Since then, thousands have been killed or injured in a tragic escalation of armed conflict and retaliation. The life of the Palestinian population was totally disrupted; on top of the acts of violence came the "closure" of residential areas, job losses and a marked deterioration in public services. From the very outset of the intifada, only one Palestinian organization proved to be prepared, even partially, to cope with the situation: the Palestine Red Crescent. And all the better prepared for the PRCS having only just completed a comprehensive VCA.

## Box 6.1 "Make people aware of the power they have"

In the early 1980s, various researchers and NGO workers recognized that the link between relief and development was, more often than not, inadequate. Mary B. Anderson, who was to conceptualize this link with Peter J. Woodrow, recalls: "I was struck to see that development work was so often interrupted by disaster. Many colleagues thought they had to stop everything in order to deal with the emergency, and would return to their development work only afterwards. And then, I often heard people in disaster-struck areas say things like 'First we had the hurricane, then the aid agencies came, and they were another disaster!' It seemed to me we were wasting opportunities offered by emergencies for initiating development work."

Part of the waste was considering beneficiaries as passive "victims", who had needs but no capacities of their own, and had to rely exclusively on external aid, which focused mainly on material aspects. This motivated Anderson and a few colleagues to find new guidelines which could link emergency and development work. The great famine in Ethiopia, in 1984, where agencies rushed, without necessarily knowing what to do and how to do it, made things easier. "People felt they needed some guidelines," says Anderson.

The guidelines were to come in 1989. After studying 30 NGO projects in various locations, Anderson and Woodrow published *Rising from the Ashes – Development Strategies in Times of Disaster*, which first introduced the notion of "capacities and vulnerabilities analysis" (CVA). CVA's main features are simple, "Number one: get to know your community. It does not take that long – the most important is to understand their capacities. Number two: never think everything can be calculated in physical, material terms." The accent here is on people's capacities, "a deeply democratic approach that makes people aware of the power they have", says Anderson. The objective is to "help people in time of need, without having them dependent on long-term aid".

A typical example from the book is drawn from Guatemala, and follows the 1976 earthquake which caused extensive damage. An NGO expert on earthquake-resistant housing visiting a severely affected community walked around the area with local people, and asked them questions on why one house had fallen down, why another was damaged, while a third was completely intact. The expert realized they had extensive knowledge of their building materials, local terrain and construction techniques. Together they designed earthquake-resistant housing that fitted the local environment better than the design the expert had brought from New York.

The CVA approach developed in parallel with another framework that favoured listening to the beneficiaries, participatory rural appraisal (PRA). Drawing from both, the International Federation developed its own methodological framework – vulnerability and capacity assessment, or VCA. ∎

In 1999, the PRCS committed itself to "advocate the need for a national disaster preparedness plan, and conduct studies to identify and establish the role and responsibilities of our society within such plans". The VCA tool was a natural choice; the same year, the International Federation had carried out a VCA pre-study, and the PRCS was already convinced of the need for "a comparison of the vulnerabilities and existing capacities to highlight shortcomings". With technical assistance from the International Federation, funding from the United Nations Children's Fund (UNICEF) for printing the assessment, and above all with a great deal of profession-alism and enthusiasm, the VCA took off. "The National Society was the engine, we merely handed them the key," observes Betts-Symonds.

The aim of the operation? To assess local communities' perceptions of the hazards most likely to occur, their needs and the resources available to prepare for and mitigate the impacts of these hazards. The approach encouraged participants to focus not only on the possibility of major disasters, but on what it is that weakens a community as it goes about its daily business. "If your definition of disaster is earthquake, people feel pow-erless," observes one of the VCA's key players, Randa Hamed. "But if you explain dis-aster through notions that are more familiar, every individual can say: 'I too can help!'"

To ensure that the various key players would collaborate from the outset – a crucial factor – they were invited to become members of a steering committee, closely mon-itoring the way the assessment was carried out. Alongside the PRCS were the International Federation, the International Committee of the Red Cross (ICRC), UNICEF and UNRWA (the UN agency responsible for Palestinian refugees). Fifteen Palestinian Authority ministries were invited, out of which four joined the commit-tee. The other ministries were interviewed in detail, as were numerous NGOs.

Nothing in the six-month assessment carried out by the Palestine Red Crescent was left to chance. PRCS social workers familiar with the target communities formed focus groups to draw out local perspectives on disaster. In one area, separate sessions were organized for men and women. PRCS staff received training in interview and group ani-mation techniques. These focus groups and interviews formed the VCA's qualitative approach, complemented by quantitative data drawn from scientific and official sources.

Communities were selected to ensure a comprehensive cross-section of Palestinian society. The 429 individuals who took part in 22 focus groups came from cities, vil-lages and refugee camps, and represented different regions of the West Bank and Gaza. The groups included traditional or religious leaders, elderly and handicapped people, and were carefully gender-balanced.

The assessment broke new ground by including 113 children and adolescents aged 6 to 19, who were asked to express in drawings their vision of disasters and ways in which they could be prevented or mitigated. "Children make up 52 per cent of our society – we couldn't just ignore them," explains Amal Shamasheh, who managed this part of the assessment. Their perception proved "often more concrete and creative than that of the adults," she adds.

chapter 6

## VCA highlights everyday disasters

The mass of data generated by the VCA was carefully analysed. The assessment high-lighted many local capacities (e.g., potential volunteers, equipment, supplies and spe-cialized staff), as well as a glaring need for training in the communities – a challenge for the PRCS. The VCA also exposed shortfalls in coordination between institutions, together with a lack of communication between communities and the authorities concerning hazard risks.

In terms of the hazards regarded as most likely to occur in the future, the results of the assessment were highly revealing. In order of importance, they were:

- lack of water;
- "events of a political nature";
- road accidents (for West Bank interviewees) and open sewers (for Gaza interviewees); and
- pollution, fires, earthquakes, poor health and epidemics.

This order of importance came as something of a surprise, even to the Red Crescent. In a stateless territory, still living essentially in a situation of occupation, one might have expected confrontations (frequent even before the intifada) to top the list. But problems related to water are a daily chore and affect everyone. Water is scarce (the region is chronically threatened by drought), expensive for the poorest Palestinian communities and frequently polluted. Nature and politics both play a part, since added to low levels of rainfall, there is competition between Israelis and Palestinians for a limited resource, most of which remains under the control of Israel.

Environmental threats – including waste water contaminating drinking water, solid waste and overflowing sewers – provide ongoing hazards. Especially in Gaza, one of the most densely populated territories in the world, with over 3,000 inhabitants per square kilometre.

The assessment also pointed to poor health as a major problem. But the picture is patchy. The population is highly educated and within it, there is a relatively high pro-portion of qualified medical staff. The same goes for hospital beds: 1.4 per thousand (more than the "golden figure" of 1.2 per thousand generally used as a benchmark for developing countries). But while a public hospital may be full to overflowing, a pri-vate establishment could be operating at only 30 per cent capacity. Similarly, one village could have three clinics managed by different bodies, while another may have no medical resources whatsoever. Health is only one illustration of how, in the absence of any genuine state structure, shortfalls and inequalities in public services abound.

Since its inception in 1994, the Palestinian Authority has developed a whole range of social services for the population, but it has been largely unable to introduce a coherent system of government. This shortfall shows when disaster strikes, be it "natural" or conflict-related. The Red Crescent, through its daily involvement within communities and its close contacts with several ministries, is well placed to appreciate the gravity of this situation. Hence the PRCS's intention to use the VCA process not only to better define their own role in disaster preparedness and response, but to advocate for the creation of a national disaster plan. This would ultimately allocate to ministries, UN agencies, non-governmental organizations (NGOs) and local authorities a clear role, without which, disaster preparedness and response become chaotic.

Finally, the VCA helped highlight what Hassan Basharat, now head of the PRCS' disaster management and coordination unit, calls "the gap within PRCS": insufficient communication and integration between departments and activities, and a need for harmonization of the work of local branches. It was, after all, only in 1994 that the organization headquartered itself in Ramallah, after half a century of exile.

The analysis of the VCA results was finalized in August 2000; 28 September 2000 marked the beginning of the intifada. "Events of a political nature" – the second most likely disaster according to the assessment – were now hazard number one, escalating, month after month, into a state of near-war, as Palestinian suicide and missile attacks were met by Israeli tank incursions, air attacks and naval bombardments. Inevitably, this modified the PRCS's priorities. Most of their emphasis would now be on responding to the new situation.

## Intifada modifies the priorities

"The intifada caught us in the middle of the process. Materially, we were not prepared. But mentally, yes", recalls Mohamad Awadeh, PRCS's deputy director of emergency medical services. "The VCA was constantly at the back of our minds," he says, "we have to plan ahead better, we have to do team work with the others, and not only with our colleagues within PRCS." Hossam Sharkawi, emergency response coordinator, adds: "The VCA acted as a key catalyst for PRCS strategic thinking and action in the direction of disaster preparedness and response. The PRCS, which traditionally reacted to various types of crisis by dispatching ambulances, moved in a direction of disaster management thinking and programming."

Of course, they *did* dispatch ambulances: the fleet was doubled in size in the course of the following months. They had no choice: the first 18 months of the intifada resulted in over 1,100 Palestinians dead and 20,000 injured, according to the *Palestine Monitor*. But the first changes were organizational: by 30 September, an operations room was working round the clock. A few days later, a disaster management and coordination unit was set up at Ramallah – an answer to the gaps in internal coordination identified during the assessment. Its tasks included constantly evaluating capacities and vulnerabili-

## Box 6.2 Experiences of VCA from Scandinavia to Africa

VCA provides a framework flexible enough to adapt to very different contexts and objectives, as the following examples show.

The **Norwegian Red Cross** (NRC) carried out a VCA in 1993. Bente MacBeath, special advisor on development, sums up the spirit of the time: "We were a western National Society, well established. Everything went well, and we took lots of things for granted. Many people felt it was the moment to check whether we were doing the right thing."

The assessment focused on identifying vulnerable groups and yielded startling results. The main target group at the time was elderly people, but it emerged that they were coping fairly well. Meanwhile, young people, single mothers and immigrants were swelling the ranks of the vulnerable. Their situation was all the more difficult since they formed a small minority in a very wealthy society. However, according to Per Christian Bjørnstad, director of NRC's national department, "The VCA undoubtedly influenced our programmes, but it was too academic and failed to find many echoes outside Red Cross circles." Another weakness was the VCA's failure to focus to any great extent on involving the community.

All this changed in 2000. In one of two surveys, 1,001 people over 18, picked randomly, were interviewed on their knowledge of and attitudes towards the social challenges posed by five vulnerable groups: children affected by poverty; young immigrants from non-western countries; lonely people; victims of violence; and people excluded from skilled society due to poor education. The second survey targeted 600 young people between

15 and 25 with the same questions. Both groups were asked if they were interested in volunteering. Results exceeded all expectations: up to 40 per cent of those surveyed stated that they were willing to take on voluntary work on behalf of one of the vulnerable groups in question. And thanks to skilful use of the media, the public debate grew to sufficiently large proportions to influence political opinion. "People are shocked to discover that in a country as wealthy as ours, thousands of children live in poverty," observes Bjørnstad.

The **Swedish Red Cross** (SRC) was pursuing the same aim. "To find out what is going on in the society," according to Thomas Kinning, SRC's advisor on capacity building. The exercise did not leave the National Society unscathed. The year-long, wide-reaching VCA, which started in 1994, focused on local chapters and provoked a painful re-examination of core assumptions. "We were shocked to find out that for 50 years we had been running some activities that no one wanted," admits Kinning.

Some SRC chapters formed a clearer picture of vulnerable groups whose existence they suspected, but who were not on beneficiary lists. These included children left on their own after school, single young mothers or unemployed people. Those chapters, while still targeting elderly people as a priority, started to assist unaccompanied children with their homework, to support single mothers and to help schools confront juvenile violence.

A fair measure of resistance had to be overcome to make the VCA a success. Many SRC chapters feared they were about to lift the lid on endless needs which they would be incapable of meeting. Sharing responsibilities with other stakeholders was an essential part of resolving this dilemma, leading the Red Cross to raise awareness among local authorities about vulnerabilities such as drug-dependency. "One of the results was to bring together various departments that were not used to collaborating, like the police and social affairs. We were really surprised when they congratulated the Red Cross for having taken the initiative!" recalls Kinning. "Little by little, our culture has started to change."

Following the first VCA, 100 headquarters staff were dispatched to district chapters to lend volunteers a hand. This in turn has led to a second VCA, taking place in 2002. "This process is vital for the Red Cross if it wants to exist in modern society," concludes Kinning.

When it launched a VCA in 1996, the **Uganda Red Cross** (URC) was facing a much more urgent situation. Two years earlier, an external evaluation had been very critical of both the URC's health programmes and its management. As a result, donors simply pulled out their support, leaving the International Federation to see what they could salvage. Apart from management problems, the National Society was burdened by a multiplicity of projects with no common denominator and little bearing on the needs of the most vulnerable. The VCA seemed to provide a way to set priorities and draw up a coherent programme, "a fundamental piece in the process of restructuring and revitalizing the National Society", as Matthias Schmale, then the International Federation's organizational development delegate in Nairobi, put it.

The survey went well, with fairly good community participation, but the project as a whole misfired, for two reasons. Firstly, an external consultant new to the process was invited to analyse the results. But the URC's middle management, who had been supportive of the VCA, was unable to accept his conclusions. Secondly, the URC's management had never felt a sense of ownership of the assessment, which it perceived as externally imposed. The picture changed only in 2000 when, under new management, the organization's national and international credibility was restored. It also began to apply the conclusions of the VCA, notably better integration of health programmes.

The new management point to several important outcomes of the VCA process: the approach is now being used by the government to develop a national disaster policy; it has helped the URC prioritize its most vulnerable branches; and it has enhanced cooperation within the National Society itself. According to Tito Kagula, acting coordinator for disaster preparedness and response, "Another result is much better cooperation with local authorities, at the district level, and with the communities. We are experiencing this right now with a project aimed at combating bubonic fever in five branches on the border with the Democratic Republic of the Congo."

Finally, points out URC's secretary general Robert Kwesiga, one key result of the VCA was that the URC only retained those roles "directly related to our mission. It was painful because people lost their jobs". Examples of activities that were dropped include distribution of essential drugs to rural health centres and providing food and shelter to street children. "You cannot address everything," argues Kwesiga. "You'd better do less, but do it well." ∎

ties, managing volunteers and integrating roles within the organization. Soon after, an emergency action plan was drafted, defining the respective roles for both Red Crescent staff and volunteers, and the agencies with which the institution works closely. The three measures had been part of the VCA recommendations, aimed at improving the PRCS's response to crisis; the intifada actually accelerated their implementation.

Other initiatives taken during the first months of the uprising had a common denominator: make the most of local resources, given severe constraints on "national" resources. For example, when tension is running high, the West Bank and Gaza can become fragmented into 62 morsels of land, isolating some communities for days or weeks. During clashes, evacuating the injured is often frustrated by military roadblocks. So, the procedure is to treat as many injuries as possible on the spot – only the most urgent cases are sent to city hospitals. To meet this demand, the PRCS redoubled its efforts to increase local capacities, from "field hospitals" (large boxes transportable by car, containing emergency supplies, including oxygen) to setting up emergency rooms in PRCS primary health care clinics, or establishing lines of communication with doctors in isolated communities. Emergency committees were set up in 21 villages or small towns, selected mainly because of their isolation.

All these measures had been recommended by the VCA. Not surprisingly so, since experience of the first intifada had shown that "closures" were one of the main problems during conflict. And the idea of employing local resources, very present in the VCA, had a scope that went beyond political conflict, and could prove useful in facing many other hazards.

## Emergency committees

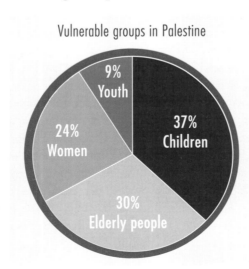

Vulnerable groups in Palestine

9% Youth
24% Women
37% Children
30% Elderly people

Figure 6.1. Palestinian communities that took part in the Palestine Red Crescent's vulnerability and capacity assessment – carried out before the intifada – consider children the most vulnerable group in society

Source: PRCS, 2000.

The emergency committees were one of the very first recommendations made by PRCS employees themselves at the outset of the VCA. Randa Hamed, at the time in charge of primary health care for the Red Crescent, explains: "Early in October 2000, we were the first institution to meet people's needs. The emergency committees were quickly set up, made up mainly of a teacher, a health professional, a member of the village council and a locally resident Red Crescent employee. They have the key to the clinic, which closes every day at 14:30 – and the list of all the medical staff in the immediate area. They intervene in the event of confrontations, but also when there are road accidents. They help assess the situation, and keep the PRCS informed in case of food or medicine shortages. The PRCS only took the initiative – the committees distributed the roles themselves, and some have been very creative."

Silwad is a small town of 9,000 inhabitants, east of Ramallah. The intifada brought serious access problems for Silwad: the shortest route to Ramallah was shut, ambulances were stopped at checkpoints, doctors were unable to reach their places of work. The community first-aid programme was extended, and a local emergency committee was created. The committee provided and coordinated health care, aid supplies, communications and environmental management. One of its members, Nadia, recalls: "One night of confrontations, we were the ones who opened the clinic and alerted the doctor until the ambulance from Ramallah arrived. We also realized that we could use cisterns to store water against drought or cuts in supply."

The idea of a committee to coordinate emergency response has also caught on in the Khan Younis camp in Gaza. There, 17 different governmental and non-governmental organizations are involved in relief, resulting in unnecessary duplication of effort. Inspired by the VCA, the PRCS established a coordination committee to bring these organizations together (see Box 6.5).

## Filling the gaps

But what about the other hazards identified during the VCA – water, public health, road accidents? Were they set aside in order to deal exclusively with the needs arising from the intifada? The answer is no. But the PRCS had to define very clear criteria, for a simple reason: everybody – the authorities as well as the communities – tends to expect too much from the Red Crescent, which enjoys an image of independence, integrity and efficiency. It was already the case before the VCA, and the tendency increased during the intifada, due to the combination of acute needs and the near-collapse of the Palestinian Authority. "From the start of the VCA, we tried to be very specific," observes the president of the PRCS, Younis Al Khatib. "We wanted to identify our role, not that of others!"

In Silwad, for example, chronic illnesses are widespread, and the lack of any system for evacuating waste water poses a genuine threat to public health. So much so that when the community was selected to take part in the VCA, social worker Abdul Ghafer Salawadeh, had to "significantly pare down the objectives, and above all reduce people's expectations, by explaining to them that the Red Crescent would see what it could do for them, but that they would have to turn to other NGOs". But, he added, "as a result of the VCA, we were committed to the community". A commitment honoured, as far as was humanly possible: in the space of a few months, the staff of the clinic, which the PRCS manages with the ministry of health, rose from six

## Box 6.3 Self-reliance in the south-west Pacific

The volcano on Manam island, off the northern coast of Papua New Guinea (PNG), is one of the most active in the region. Recorded eruptions date back to 1616. More recently, the entire island was evacuated in 1955-56. A large eruption in 1996 claimed the lives of 13 of the island's 8,500 inhabitants. Some died because they failed to evacuate before lava swept through their homes. Others died due to a lack of basic first aid. Continuous smoking and ash showers were reported throughout 2001. Yet there is no observatory tower on the island, because of a dispute between the National Volcano Observatory Authority and local landowners.

As well as volcanic eruptions, islanders are threatened by tsunami, earthquakes, landslides and drought. So, in 2001, the PNG Red Cross initiated a vulnerability and capacity assessment, with startling results. The VCA found that only 11 per cent of islanders were aware of the risks facing them, while just 6 per cent knew about resources available to deal with those risks. Over half were aware of escape routes and pickup points. But no one knew what the government's evacuation plan entailed – only one in five had even heard of its existence.

"Previously, very few of us were aware of the government evacuation plan. It's never been shared with us," says Robert Basi, a 35-year-old community leader on the island. "The Red Cross came and the VCA process opened our eyes. We now have a clear understanding of the nature and level of risks that we are facing, where these risks come from, who will be the worst affected, what is available to us to reduce the risks and what preparedness measures can be undertaken by the Manam community, collectively. And we ourselves identified all these things."

Robert's enthusiasm for the VCA process is shared by Isabell, a schoolteacher. "Women have never been asked to attend any assessment or planning session in Manam. It was a man's job only," she says. "I don't know how the Red Cross convinced our men/leaders to include women in the process. It was really good; now our men know that women can also contribute and it is quite necessary to involve us."

Based on the VCA's findings, the PNG Red Cross began a community-based self-reliance (CBSR) project. Its aim is to improve the islanders' response to disasters, by boosting faith in their own resourcefulness as a complement to government strategies. Within this process, key activities include: a **vulnerabilities and capacities matrix,** related to health, sustaining livelihoods, social relations, attitudes/behaviour and organizational structures; **community mapping** of environmental vulnerabilities, risks and hazards, including resources available in the community; a **time-and-trend analysis** of disasters in past years, listing the most severe problems and how people have coped; and a **seasonal calendar,** outlining periods of the year when the population is most exposed to hazard risks, poor health and poverty. It seems to be a popular approach. According to the headmaster of Manam's Catholic school, "It gives us new hope. Through this process, we have learned the value of self-reliance." ■

to nine, communications were improved thanks to the arrival of a fax machine, a small generator was installed to ensure that vaccinations could be safely stored and a six-month supply of medicines was created, so that the clinic could hold out in the event of a blockade.

But while health care is a core activity for the PRCS, the water issue – identified by the VCA as vulnerability number one before the intifada – is a far more difficult and political problem, beyond even the Palestinian Authority's capacity to solve. Although the PRCS could hardly secure more water for the population, they did, through the VCA, identify specific areas where they could be useful. Firstly, by providing drinking water supplies and sanitation for camps of displaced people or for mobile hospitals. During 2001, the PRCS received both equipment (e.g., mobile water units) and training from the UK-based NGO Oxfam, with whom they had strengthened their working relationship during the VCA. Secondly, the Red Crescent launched cleaning campaigns in Gaza, with the help of students, Red Crescent volunteers and the municipality. The campaigns focused not only on solid garbage, but also on the sewage system and stagnant water. According to the PRCS nurses and workers in the Jabalya camp, "The problem is to change people's behaviour. Since we started, in summer 2001, they see the difference." And, as mentioned in the case of Silwad, the local emergency committees set up thanks to a PRCS initiative are also trying to find responses to the shortage of water, such as better use of local storage capacities.

Nevertheless, as Jean McCluskey from Oxfam notes, "It is difficult for PRCS to stick to its self-imposed limits, because it is often the only one who is around." The camps for displaced people are a good example of this. Younis Al Khatib is convinced that "it is clearly not our responsibility. But during the VCA, we asked our partners to tell us what they expected of us, and then we said whether we could or couldn't perform that task. In this case, we decided we can *collaborate* with whoever is in charge – nobody seems to know who that is, by the way – for management of a camp for up to 5,000 people." This PRCS collaboration would mean providing not only water and sanitation, but also health services, support to elderly people and relief programmes with food and non-food items. When, in early 2002, hundreds of people were left homeless in Rafah, Gaza, the PRCS provided this kind of support.

This work is complemented by a huge training effort, both for PRCS staff and communities; training had been one of the most frequent and insistent demands during the VCA. Not surprisingly, the PRCS further strengthened its psychosocial units in Khan Younis and Bethlehem – a VCA recommendation which perfectly corresponded to needs arising from the intifada. And, notes Younis Al Khatib, "The telephone help-line we opened to the public is the kind of psychosocial capacity that can also be used in peace time." As for some of the other hazards highlighted by the VCA, such as road accidents and personal safety at home and work, the PRCS has started campaigns to raise public awareness about these issues.

## Getting your fingers burnt

The overall picture, however, is far from positive. The intifada interrupted the sharing of the VCA's results with the communities and other agencies. The assessment was only published in August 2001. In the meantime, some of the people who had taken part in the VCA felt let down, even when their community had benefited from initiatives resulting from the assessment.

Despite the efforts of the Red Crescent staff not to raise unrealistic expectations, frustration was at times inevitable. In Silwad, for example, a local councillor subjected the PRCS to a barrage of reproaches last November. His town was determined to have its own ambulance. But although the PRCS could provide a vehicle, Silwad had to pay the salaries. "And where do you expect me to find the money?" replied the councillor. "My own salary has not been paid this month!" Then came a more serious issue, the lack of a recycling station to treat waste water, which periodically pollutes the water table while the authorities turn a blind eye. In the absence of anyone else to confront, it was Red Crescent representatives who bore the brunt of the local councillor's ranting.

Not all the communities reacted in this way. But for Younis Al Khatib, such frustrations are quite natural. "We are not in a State, no one has clear roles. Whoever approaches the community risks getting his fingers burnt." The waste water issue, which is clearly outside the remit of the Red Crescent, begs the question: what is the point in doing a VCA if it raises issues that you cannot address?

Bassem Rimawi, deputy director of emergencies at the ministry of health, suggests one answer: "Before taking any action, we have to fully understand the situation: what resources do we have, what do we need? No one had so far carried out this research in our communities." To which you could add that, when the VCA does identify issues outside your remit, you can either try to find other organizations to do the job, or advocate for changes to the system.

## Leading by example

"Advocacy takes time and hard work," says Younis Al Khatib, who is looking forward to being able to approach the ministries to promote much-needed rules and regulations. Above all, he wants to encourage them to advance the drafting and adoption of a national disaster plan. But this gets more difficult by the day. Territorial fragmentation prevents the representatives of the various ministries, scattered here and there,

## Box 6.4 Key principles for a successful VCA

- **Driven by those at risk.** It is vital that the communities and aid organizations in the area at risk feel a sense of ownership in all stages of the assessment. If they get the impression that the VCA is being imposed from outside, there is a very real risk of failure. Using consultants can work, but only provided they act as *facilitators* and cooperate closely with participating communities and organizations. The involvement of staff from aid organizations in the area is indispensable.

- **Full commitment.** A VCA is not a risk-free exercise. Embarking on a VCA means recognizing and agreeing to accept both the additional workload it will represent and its results, which may be unexpected and challenging.

- **Access to available resources.** It is essential to start by collecting data already available (statistical and otherwise), and not to reinvent the wheel. Similarly, it is useful to team up for the assessment with people who are highly skilled in the field to which the VCA relates.

- **Good training and preparation.** The objectives of the assessment must be clear from the start. Survey methods used vary according to local circumstances. Some VCAs may use a questionnaire, while others may opt for a more open approach. What is important is that the individuals due to carry out the survey are familiar with the target community, enjoy their trust, and receive good training. Timing may also be important; in communities prone to drought or extreme cold, for example, the season in which the survey is carried out may highlight vulnera-

bilities which may not be apparent at other times of year.

- **Participation of interested partners.** The relevant authorities, both local and national, NGOs or international organizations must all come on board from the very outset of the process. This will ensure that they feel involved in the results of the assessment and that they are fully in the picture when taking part in any subsequent projects.

- **Participation of communities.** As for partner organizations, the communities concerned should not be merely consulted, but actually *involved* in the process. This pays off in two ways: the VCA becomes in itself a community-learning process; and its results are more likely to be accepted by those concerned.

- **Good communication with all involved.** The organization carrying out the VCA must share its findings on a regular basis. It is vital that the analysis phase in particular – following data collection – is handled with great care. It should be followed by feedback to the organization's staff, partners and communities involved.

- **VCA should be adopted as a way of working.** A good VCA is not an interlude in the life of an organization, but a springboard to a different way of working, a new way of seeing things. In particular, partnerships (with other organizations and target communities) should become the rule, as should a constant reappraisal of the facts. It is therefore a continuous process, based on the assumption that circumstances (capacities, vulnerabilities, etc.) are in a constant state of flux. ■

from meeting – many leaders reunited in a single place to debate a national disaster plan could become a military target, according to those concerned.

Nevertheless, the example of ambulances provides some positive results of PRCS's advocacy. The white vehicles bearing the telephone number 101, always visible in accidents and areas of confrontation, are the PRCS's calling card. Thanks to regular support from the ICRC and donor National Societies, the vehicles are in excellent condition and well-equipped, and the ambulance teams are highly trained. Better still, the Red Crescent is officially mandated by the ministry of health for everything which relates to pre-hospital emergencies.

Yet virtually everyone manages their own ambulance service. Many villages boast of having a "white vehicle" funded by some generous donor, giving them the illusion of self-sufficiency. But in practice, "it's often little more than a taxi, and the driver, having received absolutely no training, is likely to cause more damage by handling the injured", observes Fayez Djibril, deputy director of the Red Crescent emergency medical services at Gaza. Many agencies also manage an ambulance service – UNRWA, military medical services, civil defence and the ministry of health. The PRCS has promoted common rules for all, with some success; not only does it provide training to non-PRCS ambulance staff, but it has worked with the ministries of transportation and health on strict criteria to provide licences (valid for 12 months) for ambulance staff, drivers and even vehicles. Enforcement of the new rules, however, remains an issue.

Meanwhile in Nablus, the governor has, of his own initiative, been promoting teamwork in various fields (e.g., health, social services, support for families of those killed) since the beginning of the intifada. An emergency room, on the model of the Red Crescent, centralizes information and coordinates action. An emergency committee can summon all stakeholders together in serious situations. The governor has asked the Red Crescent to provide training in disaster preparedness and response. "Cooperation between stakeholders is still far from perfect, but meanwhile, we are the only ones in the West Bank to have such an organization," says the governor.

## People-centred process

The story of the PRCS is complex and paradoxical. The territory it works in does not exist as a state and it is plagued by armed conflict. Clearly, the "post-VCA" story would have been different without the intifada. As Younis Al Khatib, puts it, "The VCA identified a wish-list. The intifada refined it, leading to more sustainable programmes." The intifada undoubtedly accelerated the implementation of many ideas put forward during the VCA. But it has also unbalanced the PRCS's activities, tipping the scales towards conflict-related emergency response, and indefinitely postponing wider activities such as promoting a national disaster plan.

Was a VCA even necessary to come up with the idea of an operations room once the uprising had started? Wouldn't common sense have led the PRCS to the same conclusion? The answer is certainly yes. The same goes for the idea of creating local emergency committees, or strengthening the internal organization of the PRCS. But without the VCA, these would not all have happened in parallel, nor so fast. The freshly completed assessment provided the PRCS with a clear picture, a mental framework that enabled it to make quick choices in a particularly chaotic situation.

The experience of the PRCS shows that any organization undertaking a VCA must be willing to put their backs into it, to accept the results (however challenging these may be) and to put its recommendations into action. A unique aspect of the process, points out Graham Betts-Symonds, is that "you can never be quite sure what you will find". The assessment could last months or years. It could focus on communities, or more on the implementing agency itself. Its priority may be to highlight capacities, or to identify vulnerable groups more accurately. Whatever the particular aims of specific VCAs, they share three overriding principles:

- **VCA puts people first.** Rather than relying simply on technical systems to determine hazard risk, the VCA approach reveals the risks which vulnerable people perceive to be most threatening. And unlike traditional needs-assessments, VCA concentrates as much on the capacities of exposed communities as on their needs and vulnerabilities. People-centred assessment ensures that actions taken by authorities, aid organizations and communities themselves will be more relevant to real needs and available resources.
- **VCA is a process, not a product.** VCA does not aim to provide a "snapshot" situation report. It is a learning process from the start, which grows into a long-term way of assessing the operating environment. For the Palestine Red Crescent, the VCA has become a continuous process – a way of life, underpinning their daily work. Even more so, now that they are living in a state of continuous crisis, where decisions have to be quick and accurate.
- **VCA involves all players from the outset.** For the Palestinians, the VCA provided the catalyst for a closer working relationship with key actors. This greatly improved cooperation between the PRCS, NGOs and authorities from the start of the uprising. Drawing in the full range of players from the outset is the only way to create ownership of the assessment process, and of the programmes which follow.

The Palestinian context may seem uniquely difficult, but the operating environment of humanitarian crisis aggravated by weak state structures is shared by many developing nations. In such environments, VCA will inevitably highlight major risks and shortfalls, and raise expectations that something will be done about them. The range of challenges that a VCA could raise risks paralysing aid organizations and authorities into inaction, or prompting them to take on a multitude of unsustainable pro-

## Box 6.5 Hospital of hope for victims of intifada

A piercing shriek rings out; a shriek so full of rage, so full of despair, that it makes the visitor jump. It comes from a group of young boys playing in the hospital corridors. Jean Calder calmly explains, "That's a little boy of eight, one of our most serious cases. He is suffering from emotional trauma having witnessed scenes of violence against his family." Such violence is commonplace in the Gaza Strip, for years one of the most overpopulated and destitute spots on earth, home to a million Palestinians and hermetically sealed off from the outside world.

"People live here without hope," comments Aziza, a nurse working for the Palestinian Red Crescent at the El Amal hospital, which serves the Khan Younis refugee camp. "Just how can we help them, these people who no longer have any work, who have to try and get through the winter, whose sewers are overflowing, and who've seen nothing good come out of the past few years?"

And yet, Aziza, Calder (who left her native Australia 20 years ago to work with the PRCS) and so many of their colleagues are doing what they can to restore a little hope. At the El Amal hospital (whose name signifies "hope"), built by the PRCS, a project has been set up for over 300 injury victims of the intifada, most of them adolescents and their families. "We quickly realized that although the physical treatment was adequate, there were enormous psychological and social problems," explains Calder. "Above all people need to

regain a little of their self-respect." Apart from providing psychosocial support, the programme includes both educational and vocational training (such as sports, music, handicraft activities, computer skills or English classes) and financial help with rebuilding houses that have been demolished. For those, and they are legion, who prefer to suffer in silence, unable to leave their own homes, therapists and social workers carry out an increasing number of home visits.

Calder, who took an active part in the Palestinian VCA, stresses the enormous need for coordination where humanitarian aid is concerned. "In Khan Younis alone, there are 17 government and non-government organizations involved, and we discovered that sometimes several of them were coming to the aid of the same patient, unbeknown to the others," she says. "That's why the PRCS took the initiative of introducing a committee bringing together all those involved, which was in fact one of the VCA's recommendations, and they all use the same patient referral form now."

"In this country, we have lots of systems," comments Mohamed Awadeh, PRCS's deputy director of emergency medical services. "But not one system, and no common language. We may intersect at times, but we do not really meet." In many parts of the Palestinian territories, it sometimes seems like no one is pulling in the same direction. In Khan Younis, at least, they are trying. ■

grammes. It is perhaps here that the VCA process can play a key part – in building a constituency of communities, aid organizations and local authorities, aware both of their needs and capacities. This constituency can, in turn, work together to tackle the challenges which may seem insuperable to a single agency, ministry or community.

*Iolande Jacquemet, an independent writer based in Geneva, was principal contributor to this chapter and to Boxes 6.1, 6.2, 6.4 and 6.5. Latifur Rahman, the International Federation's disaster preparedness delegate in Papua New Guinea, contributed Box 6.3.*

## Sources and further information

Anderson, Mary B. and Woodrow, Peter J. *Rising from the Ashes. Development Strategies in Times of Disaster.* Boulder, Co.: Lynne Rienner Publishers, Inc., 1998.

Christoplos, Ian, Liljelund, Anna and Mitchell, John. "Re-framing Risk: the Changing Context of Disaster Mitigation and Preparedness" in *Disasters,* 2001, Vol. 25, no. 3, pp. 185-198.

International Federation of Red Cross and Red Crescent Societies. *Vulnerability and capacity assessment.* Geneva: International Federation, 1999.

Palestine Red Crescent Society. *Vulnerability and Capacity Assessment.* Al-Bireh: PRCS, 2000.

Palestine Red Crescent Society. *Emergency Readiness Plan.* Al-Bireh: PRCS, 2001.

Norwegian Red Cross and ECON (Centre for Economic Analysis). *Social Pulse 2001. Flaws in the Welfare Society.* 2001.

Awwad, Dr. Elia. *Report on Trauma and Loss Among Palestinians.* PRCS, October 2001.

## Web sites

Collaborative for Development Action **http://www.cdainc.com/cda-home.htm**
International Federation **http://www.ifrc.org/what/dp/VCApalestine.asp**
Norwegian Red Cross **http://www.redcross.no/**
Palestine Red Crescent Society **http://www.palestinercs.org**
Swedish Red Cross **http://www.redcross.se/**

chapter 7

Section Two

**Tracking
the system**

# Accountability: a question of rights and duties

*"The very high mortality resulting from epidemics of cholera and dysentery among Rwandan refugees in Goma, Zaire, in 1994 was a turning point for humanitarian aid organizations. Wide-scale clinical mismanagement of cholera by inexperienced relief workers was reported in both the scientific literature and in the news media. Prior to this time, many humanitarian organizations had recognized the need to improve professional standards within their own organizations to improve the effectiveness of their humanitarian response. Events in Goma, however, increased the impetus to improve accountability within the humanitarian system as a whole."*

Improving standards in international humanitarian response:
the Sphere project and beyond

Over the years, humanitarian practitioners and evaluators have consistently highlighted a number of problems frequently encountered in humanitarian operations, including absence of professionalism, poor management, problematic funding policies and practices, absence of coordination, lack of humanitarian access and military targeting of civilian populations and relief workers.

Governments, armed forces and donors bear primary responsibility for some of these problems, such as lack of humanitarian access or attacks on civilian populations. Other problems, however, directly challenge humanitarian actors, such as United Nations (UN) agencies, the International Red Cross and Red Crescent Movement and international and national non-governmental organizations (NGOs).

Addressing these problems involves unpacking the nature and extent of humanitarian responsibilities and placing a greater focus on accountability, both as a pre-eminent humanitarian principle and as an institutionalized practice. This chapter will analyse why accountability is important, what it means in practice, who is responsible for what, and how to operationalize accountability at both field and headquarters levels.

## Why does accountability matter?

As early as 1983, Fred Cuny argued that:

*"Almost from the very beginning of intervention, the first question that arises is one of accountability. Ask staff members of almost any relief organisation to whom they are accountable, and they will probably reply to their home office, to their accountants and, almost always, to the agency's donors. Some may respond that they are also accountable to*

Photo opposite page: Humanitarian actors exercise power over the lives of crisis-affected people. With power comes responsibility towards the individuals they are serving.

Christopher Black/ International Federation, Guinea 2001.

chapter 7

*the government of the country in which they are working. A few may even say that they are accountable to their counterpart agency in the country. But where in this list is the victim? If this is truly the helping relationship that most agencies would like to achieve, why has the victim been left out? … The concept of accountability to the victims of a disaster is a concept long overdue in relief practice. Without accountability, programmes inevitably become paternalistic in nature or end up serving the needs of the donors and the agencies rather than the needs of the victims."*

The call for greater accountability constitutes one of the key features in the humanitarian landscape of the past two decades. For example, the importance of human rights standards in assessing actions (and inactions) by states, corporations and other non-state actors has increased. Civil society, often the originator of calls for greater accountability, has not escaped the accountability spotlight either.

But why is accountability important in the context of humanitarian action? Put simply: humanitarian actors exercise real power over the lives of crisis-affected individuals and communities. Power to decide who receives aid and who does not; what will be given, when and where. Power to determine where people must go and when, what they will eat, what clothes and shelter they will have, and how much private or social space they will enjoy. An extract from the joint evaluation of emergency assistance to Rwanda, published in 1996, illustrates the point:

*"One agency (Operation Blessing) sent six medical teams of 15 people each after an appeal for doctors on a TV programme produced by its parent organization, Christian Broadcasting Operation. Staff were rotated every two weeks and most had no previous experience in Africa. The teams undertook inappropriate activities, such as setting up IV infusions in patients' shelters and leaving them unsupervised, and because of Operation Blessing's reliance on donated drugs, it had only one antibiotic available for all prescriptions. Americares, another US-based NGO, airlifted 10,000 cases of Gatorade, a popular sport drink, which cannot be considered a beneficial treatment for cholera."*

Even worse than this kind of negligence are disturbing reports of aid agency staff in west Africa abusing their positions of power by demanding sex in return for aid. On 27 February 2002, the UN High Commissioner for Refugees (UNHCR) and Save the Children UK released a report on the sexual exploitation of refugee children in Liberia, Guinea and Sierra Leone. Agency workers from international and local NGOs, as well as UN agencies, were reportedly the most frequent sex exploiters of girls under 18, often using the very humanitarian aid and services intended to benefit the refugee population as a tool of exploitation. According to the report, refugees had tried to send written said through staff but the information had been held back. Children said that they were harassed, labelled or denied services when they tried to complain. Agencies are currently working to investigate the allegations.

## Box 7.1 Rights-based programming in Sierra Leone

In Sierra Leone, the NGO CARE is exploring whether, through humanitarian and recovery assistance, it can address, at local level, some of the root causes of recurrent conflict. In a number of remote villages, CARE is piloting a rights-based approach to food security that seeks to address questions of governance among its target population. CARE also aims to analyse how it relates to its beneficiaries through staff attitudes and programme management.

When asked for a forum where the causes of conflict could be discussed, the villagers suggested a traditional event. They called it the "Peace and Rights Day", and it generated great interest. CARE ensured that women and young people could discuss their views in separate focus groups, and that their views were heard and recorded.

Through this forum, villagers openly recognized that the local authority structures, through which CARE (like other agencies) used to work, were not truly representative of the community. They agreed that aid inputs were not always equitably distributed between villages, between families or even within extended families. So they decided to elect committees of youth, women and elders to act as mechanisms for representation, regulation and enforcement. CARE agreed to support the committees with training in committee management, leadership, basic knowledge of the law and legal procedures, with the aim of fostering more just, transparent and accountable local governance.

Simultaneously, CARE has made significant changes to its seed programme. Rather than providing another round of imported seed in bulk to heads of farming families, every individual with the will and possibility to farm was asked to come forward and be registered. They were asked if they would like any particular seed variety that they had valued highly before the war, but which was no longer locally available. CARE staff then scoured Sierra Leone to find and purchase the seed variety locally, which was then provided to the specific individual, with her or his name tag on it.

This change in seed-distribution strategy not only serves to underline the message that every individual has the right to food security. It also dramatically demonstrates the desire and possibility of a humanitarian agency to respond to individually expressed preferences. Other outcomes of this approach may include not only the re-introduction of greater seed variety into the area, but also a resumption of the "seed-fairs" which bring people together in acts of exchange.

CARE staff continue to visit the villages, and actively seek feedback from a range of individuals on the quality of the seed they have received, and their views on the older and newer methods of distribution. The staff encourage critical comments from people, seek positive suggestions and engage locals in discussions about their rights and responsibilities. Through supporting events (Peace and Rights Days), programme design (tailor-made seed distribution) and the nature of their relationship with beneficiaries (respect for everyone's opinion, transparency about programme resources and constraints), CARE's project staff are improving the quality of aid provision and helping encourage local people to promote better governance. ■

Building a culture and practice of accountability within the humanitarian sector requires first acknowledging that in providing relief, humanitarian actors often acquire great power. And with power comes responsibility. The second step consists in recognizing that humanitarian action is ultimately not about logistics and commodities, but about individuals – and individuals with rights.

## What is accountability?

Within the humanitarian sector, experiences from Goma in the mid-1990s, and from other humanitarian operations have prompted agencies to look more closely at the issues of quality, accountability and legitimacy. As the joint evaluation of emergency assistance to Rwanda points out:

*"It is unacceptable that an NGO with little or no relevant experience is able to send personnel to a humanitarian relief operation and engage in activities that discredit or undermine the overall efforts; provide unacceptably poor standards of service and care to their beneficiaries; and then leave without any recourse."*

So accountability requires recognizing and implementing two interdependent principles:
- Individuals, organizations and states should account for their actions and inactions and be held responsible for them.
- Individuals, organizations and states should be able to report safely and legitimately concerns, complaints and abuses, and seek redress where appropriate.

Accountability aims at ensuring that the power of humanitarian actors is exercised within a framework of fairness, respect and justice. Crisis-affected people are not a number on a ration card or a shadow in a queue – they are individuals, with a right to be informed and consulted, to participate in decisions affecting their lives, to raise concerns and complaints, and to get answers. Accountability mechanisms should be applied in policy and practice at all stages of the actor-beneficiary relationship – whether during disaster preparedness, emergency response or reconstruction. Humanitarian accountability can be summarized as follows:
- **Who is accountable?** Duty-bearers with a responsibility towards crisis-affected populations. These include governments, armed forces, NGOs, the Red Cross Red Crescent and UN agencies.
- **To whom?** Duty-bearers are accountable, first and foremost, to populations and individuals affected by disaster and conflict. They are also accountable to their donors.
- **For what?** To meet responsibilities as defined by agreed benchmarks such as legal standards, ethical principles and professional, agency or interagency codes, standards or guidelines.

## Box 7.2 Concerns and complaints

Crisis-affected people are grateful for the assistance they receive, but they often also have valid concerns and complaints. These can generate positive changes in the nature of aid services and the ways they are delivered. Frequent concerns and complaints, voiced by crisis-affected people, relate to:

- **Entitlements.** Unfair targeting or allocation criteria; diversion of assistance; abuse of power by aid-providers; lack of information about relief or compensation programmes.
- **Quality of assistance.** Untimely; inappropriate; inadequate; inaccessible (e.g., for the elderly, disabled); not meaningful (e.g., food-for-work).
- **Physical and legal protection.** This can be a higher priority for affected people than material assistance.

- **Dignity.** Labelling people as "victims", "poor", "displaced", "dependent".
- **Local capacities.** Not recognized or employed; undermined because relevant information not given in timely way.
- **Distrust.** Between aid provider and recipient. Arrogance, even racism, of aid workers; too many assessments; perceived broken promises; unaddressed grievances.
- **Unintended impacts.** The way aid is provided can aggravate social conflict, undermine livelihoods and encourage "dependency".
- **Root causes.** No fundamental improvements likely unless root causes are addressed and durable solutions found. ■

- **How?** Through establishing mechanisms of accountability at field, headquarters and interagency levels. These include setting standards and indicators, monitoring activities, investigating complaints, reporting to stakeholders and identifying duty-holders.
- **For what outcomes?** Changes in programmes and operations, sanctions, recognition, awards and redress.

## Accountability and quality

Over the last decade, quality assurance has become the other key debate within the sector, alongside accountability. The debates and initiatives around both issues have often intersected. A recent analysis by English academic John Borton shows that two main elements of the broader quality debate are relevant to humanitarian work:

- The emphasis on measurable characteristics and specifications.
- The central role of the consumer or beneficiary.

Borton suggests that quality improvements can be encouraged within the humanitarian system by refining the European Foundation for Quality Management (EFQM)

Excellence Model. EFQM was founded in 1998 by the chairmen of 14 major European companies with the objective of stimulating corporate interest in Total Quality Management. EFQM defines quality as "the product of a combination of customer satisfaction, people satisfaction and impact on society ('Results') which are achieved through leadership driving policy, strategy, people management, partnerships and resources and processes ('Enablers'), leading ultimately to organisational excellence."

Thus, quality assurance constitutes a managerial tool and objective which seeks to ensure shareholder and consumer satisfaction. Accountability implies an emphasis on the activities and processes through which quality will be monitored and ensured. It differs from quality assurance in its focus on the legal and ethical *responsibilities* of duty-bearers and on the *rights* of affected populations – including the right to raise complaints and seek redress.

Nevertheless, accountability and quality assurance share several common features, including the development of adequate policies and processes, the need to monitor activities and impact, the central role of an organization's leadership and the emphasis on stakeholder satisfaction.

Initiatives since the mid-1990s to improve the quality of humanitarian aid have included the *Code of Conduct for the International Red Cross and Red Crescent Movement and NGOs in Disaster Response (Code of Conduct)*, the *Humanitarian Charter and Minimum Standards in Disaster Response* (Sphere project), the Ombudsman project, the Active Learning Network for Accountability and Practice (ALNAP), the Local Capacities for Peace Project ("Do No Harm") and People in Aid.

More recently, the Humanitarian Accountability Project (HAP) was launched in Geneva in 2001, in response to concerns about the lack of accountability within the humanitarian sector towards crisis-affected populations. The HAP is working at field, organizational and sector-wide levels to identify, test and advocate for accountability mechanisms.

These various initiatives seek to address some of the necessary characteristics for organizational excellence, as defined above, for example, development of principles guiding humanitarian work; minimum qualitative and quantitative standards; analytical tools to consider the impact of aid on conflicts; better management of humanitarian relief staff, and so on. They also seek to contribute to founding an accountable humanitarian sector by insisting on humanitarian responsibilities. Guided by standards enshrined in human rights legislation, refugee law, international humanitarian law (IHL) and International Labour Organization (ILO) conventions, they aim to adopt, explicitly or implicitly, a rights-based – rather than a purely needs-based – approach that includes a focus on the responsibilities of humanitarian actors.

## Box 7.3 Right to information

About a month after the devastating earthquake which shook the Indian state of Gujarat in January 2001, *Abhiyan*, a grouping of over 20 local voluntary organizations in Kachchh district, decided to stop its relief work and start concentrating on recovery.

*Abhiyan* identified a strong demand among crisis-affected people for information, first of all about the policies of the Gujarat state authorities on rehousing and drought relief. At issue were insufficient, unclear and incomplete information, including confusion over who was entitled to compensation, and disagreement over assessments conducted by government-appointed surveyors.

*Abhiyan* acknowledged that people had a right to information. Because so many other agencies continued doing relief and reconstruction work, its members chose to transform their relief distribution points into information and advice centres. Their functions were to:

■ provide relevant, complete and accurate information to affected people, covering government policies, procedures, funds available and entitlement criteria;

■ develop a communications strategy and materials to disseminate this information in a form accessible and understandable to all;

■ work with co-residential groups to develop reconstruction plans, and to decide how best to use available funds;

■ act as a "help-desk" where people could seek advice or bring grievances, and which could encourage responses from aid providers to these grievances; and

■ be a major meeting point for affected people and aid providers.

The underlying goal for this initiative was, in *Abhiyan*'s words, "to ensure that people develop the ability to negotiate from a position of strength. Making informed choices would be possible if people have access to information and space to analyse and discuss it with the community." ■

# Duty-bearers and their responsibilities

A variety of actors are involved in humanitarian crises. They range from national governments and their military forces to donor governments, aid agencies and affected populations. All those working in humanitarian relief should be accountable and they should all be accountable to crisis-affected populations. This principle has already been well articulated, through the *Code of Conduct* and various individual agency mission statements.

But humanitarian actors are not all accountable for the same things. It is important to distinguish between different types and levels of responsibility. For example, the duties of a national government towards affected populations and the standards by which it has to be held accountable differ from those of humanitarian NGOs or UN agencies.

## Box 7.4 Popular involvement becomes political in Nicaragua

Following Hurricane Mitch in October 1998, 21 Nicaraguan networks, representing some 350 national NGOs, social movements, unions, collectives and federations, came together in the Civil Coordinator for Emergency and Reconstruction platform (CCER). Coordination of emergency relief was an immediate objective. But a wider goal was to transform reconstruction efforts into an opportunity to address the underlying causes of vulnerability in Nicaragua. So CCER formulated positive proposals, strengthening their legitimacy through extensive consultations in thematic commissions and through population-wide "social audits".

The first social audit was conducted in February 1999. It used quantitative and qualitative approaches, targeting over 10,500 households and almost 300 community leaders throughout the country. Two more audits followed. Their primary aims were to: (a) **gather baseline data** on the situation of crisis-affected communities, in order to prioritize and adjust aid programming; (b) **define indicators** for measuring: the needs and priorities of women, young people and children; how affected people perceive the aid effort; and the progress of affected people during reconstruction and development; and (c) **establish a basis for dialogue** and mutual accountability between CCER, other social actors, the Nicaraguan government and the international aid community.

CCER acknowledges that this effort to place people at the centre of the discussion constitutes a political process, because it seeks to integrate the views of communities into the planning and evaluation process, to build alliances that promote and broaden citizens' participation in the reconstruction process, and to strengthen the impact of civil society on public policy through the formulation of its own proposals.

Among the topics tackled were the efficiency and equity of aid distribution, aid's transparency and coordination, and whether people felt their opinions had been taken into account. Findings were presented in a draft report, which was shared with and discussed by the communities, civil society representatives and mayors from the regions surveyed.

The second social audit also presented its results graphically, in the form of "social maps", which visualize respondents' answers by geographical area. It is even possible to superimpose, and thereby correlate, respondents' feedback with, for example, maps of the geographical distribution of damage done by Mitch, maps of endemic poverty and political maps showing which party controls which districts.

Not surprisingly, CCER's activities and its successful lobbying of international aid donors led to tensions with the Nicaraguan government. Concerns about transparency and corruption meant that more foreign aid was flowing through NGOs than through government channels. Audits revealed a negative impression among the population of the government's performance in the reconstruction effort. And CCER criticized the government's strategy for poverty reduction, both in terms of the evidence presented on the situation in the country, and the capacity of the proposed policies to address the magnitude of the problem.

Tensions became more overt when, in February 2000, the government cancelled a planned meeting with the World Bank-led Consultative Group. In response, CCER organized a national civil society forum to evaluate the progress of reconstruction. National media carried attacks on CCER's official spokesperson. And members of both the government and the opposition started challenging the legitimacy of this "civil society" voice, which operated outside established party political structures.

CCER's experience so far highlights some fundamental issues. Civil society can mobilize itself and effectively lobby at international levels to get policy demands heard but, at the same time, it cannot replace all government functions. Each side will therefore have to acknowledge its limitations, and show itself open to responsible debate and collaboration. ■

## Responsibilities of states and armed groups

States remain the primary duty-bearers with regard to respecting human rights. Along with armed groups, they are obliged to respect humanitarian standards as defined by the Geneva Conventions and to prevent their violation. Moreover, states may be held to account for their failure to take reasonable steps to prevent or respond to an abuse committed by a *non*-state actor, including not only those to which states have delegated activities, but also private individuals or corporations.

Under human rights standards, states have an obligation to secure universal respect for and observance of human rights. For instance, the International Covenant on Economic, Social and Cultural Rights (ICESCR) obliges states to "take steps individually and through international assistance and co-operation…to the maximum of [their] available resources" to realize the right to food, clothing, housing and health.

Where and when states are unable to assist crisis-affected populations, it has been well argued and defended that they have an obligation to seek assistance from others – and therefore to allow access for humanitarian purposes. The principle of humanitarian access should apply both in conflict and non-conflict situations.

In 1999, this principle of access was strengthened, within the ambit of human rights law, by a statement from the Committee on Economic, Social and Cultural Rights, the body responsible for interpreting and monitoring the implementation of ICESCR. The committee insisted, in a general comment on the right to food, that the "obligation to respect existing access to adequate food supplies requires States parties not to take any measures that result in preventing such access" and that "violations of the right to food can occur through… the prevention of access to humanitarian food aid in internal conflicts or emergency situations".

IHL, as enshrined in the 1949 Geneva Conventions and their 1977 Additional Protocols, also recognizes the right of victims to receive assistance, in non-international as well as international conflicts. The obligation to allow free passage of relief is made conditional upon: (a) checks to ensure that the contents are distributed exclusively to those for whom it is intended; and (b) a formal request to, and reply by, parties to the conflict.

Meanwhile, the principle of access is central to many humanitarian texts. For example, the *Code of Conduct* states: "The right to receive humanitarian assistance, and to offer it, is a fundamental humanitarian principle which should be enjoyed by all citizens of all countries… Hence the need for unimpeded access to affected populations, which is of fundamental importance in exercising that responsibility."

Furthermore, the UN, in a recent resolution on strengthening the coordination of emergency assistance, said, "States whose populations are in need of humanitarian assistance are called upon to facilitate the work of these organizations in implementing humanitarian assistance, in particular the supply of food, medicines, shelter and health care, for which access to victims is essential."

And humanitarian access is spelled out in Principle 25 of the Guiding Principles on Internal Displacement, formulated by the UN and other partners, according to which: "The primary duty and responsibility for providing humanitarian assistance to internally displaced persons lies with national authorities" and "International humanitarian organizations have the right to offer their services in support of the internally displaced. Such an offer shall not be regarded as an unfriendly act or an interference in a State's internal affairs and shall be considered in good faith. Consent thereto shall not be arbitrarily withheld, particularly when authorities concerned are unable or unwilling to provide the required humanitarian assistance."

Yet, despite the consistency of international law and ethics in promoting the meaning and applicability of the principle of humanitarian access, the practice is sadly different. Governments and armed groups too often violate this principle, in the name of "national sovereignty" or war efforts. And humanitarian workers are, too often, the deliberate targets of intimidations or killings. In Afghanistan, Chechnya, Colombia, the Democratic People's Republic of Korea and Sudan, for example, restrictions have been imposed on the provision of humanitarian aid or humanitarian workers have been targeted.

Finally there remains an imbalance in legal provision between those affected by conflict and those affected by "natural" disaster. The rights of the former are protected under IHL and the Geneva Conventions, while the latter enjoy far less legal protection. The idea of systematizing a body of International Disaster Response Law (IDRL), detailed in the *World Disasters Report 2000,* is one attempt to redress this imbalance (see Chapter 9, Box 9.1).

## Box 7.5 Right to registration

Hundreds of thousands of Chechens have become displaced by the second Russo-Chechen war that broke out in September 1999. One of the few international aid organizations working in the northern Caucasus is the Danish Refugee Council/Danish People's Aid (DRC/DPA). After noticing early on that a number of displaced persons had registered more than once, DRC/DPA started a series of surveys to develop a fair and efficient system for aid distribution. As a result, there is now a database in which almost a million Chechens, residing in Chechnya, Ingushetia and Daghestan, are registered.

More importantly, DRC/DPA has created a number of "complaints centres" (three in Ingushetia and six in Chechnya), which complement a consultation centre run by UNHCR. Chechen people can approach these centres if they have not been registered, or have become de-registered (for example, as a result of not being present at aid distributions). They may not have received aid to which they think they are entitled, they may want to ask about a change in the aid offered in line with changes in their circumstances, or they may wish to receive their aid elsewhere if they have moved house. Whatever their concern or complaint, they can present it in person or in writing, and DRC/DPA or UNHCR lawyers will examine their case and report back to them on the outcome. ■

# Responsibilities of civilian actors

States, however, are not the only actors with responsibilities under human rights legislation and IHL. Non-state actors are also directly accountable for their actions, both nationally and internationally. The Universal Declaration of Human Rights (UDHR) calls on "every individual and every organ of society" (including, therefore, civilian humanitarian NGOs) to play their part in securing universal observance of human rights. The Geneva Conventions include articles related to the rights but also the *obligations* of the International Committee of the Red Cross (ICRC), National Red Cross and Red Crescent Societies, and any personnel participating in relief actions. And increasingly, humanitarian actors working with refugees see their role as fulfilling refugees' rights to protection.

The development of the *Code of Conduct,* Sphere's *Humanitarian Charter*, agency and interagency professional standards, and the mission statements of many individual humanitarian organizations indicates the commitment of humanitarian actors not only to integrate their actions within a human rights context but also to define their roles in terms of human rights responsibilities. These initiatives clearly reveal a sector seeking to move away from a simple ethic of charity towards one grounded in obligations and rights. Paul O'Brien, CARE's policy advisor for Africa, argued in a recent article about promoting rights and responsibilities, that:

*"I have come to believe that a rights-based approach in its rawest form, means asking and answering three questions: (1) How can I treat the people I serve as responsible actors? (2) How can I take full responsibility for the impact of my work? (3) How do I ensure others fulfil their human responsibilities?*

*"I believe that to do rights-based programming is to ask and answer these questions in every programme, all the time. These questions need to be asked both individually and institutionally. Nothing more is required to do rights-based programming, but if we do less, I believe we are failing our human responsibilities.*

*"How does this answer the thoughtful sceptic who sees no real difference between solid needs-based programming and this new 'snake oil'? The best response I can give is to look at how 'rights' are different from 'needs'. In my view, there are two critical distinctions between rights and needs. First, rights always trigger responsibilities whereas needs don't, and second, rights inherently imply standards against which responsibilities can be measured whereas needs don't."*

## Multiplication of principles, codes and standards

Over the last decade, the humanitarian sector has witnessed the mushrooming of guidelines, codes and benchmarks at field, organizational and interagency levels. Some of these establish general principles, including rights-based principles, governing humanitarian action (e.g., the *Code of Conduct*). Others establish specific qualitative or quantitative benchmarks at sectoral level (e.g., Sphere's technical minimum standards).

Many organizations have developed or strengthened their own internal management guidelines and professional standards; a large number have also developed and adopted interagency standards. At field level, a number of individual organizations have begun to embrace accountability through the adoption of rights-based programming (see Box 7.1).

But does this multiplication of principles, codes and standards complicate matters unnecessarily? Does it thwart attempts to offer a single baseline against which to ground and assess humanitarian operations systematically?

The answer is no, for two reasons. The first reason is that the development of these codes or standards has been very much driven by contextual and operational concerns. So they are based in the reality of humanitarian work, which often requires negotiating between a variety of different principles and actors (many of whom consistently violate their obligations). As such, these texts offer a practical,

## Box 7.6 Right to appropriate assistance with dignity

During the Balkan wars, the International Federation ran various programmes to distribute relief items. During a 1994 training course, it became clear that logistics staff saw "delivery" of relief goods as their primary role. In order to improve the quality of aid provision, the International Federation's desk officer encouraged all delegates to spend at least one day a month canvassing the opinions of a random sample of beneficiaries about the timeliness, appropriateness and content of the relief packages. This initiative led to intensive discussions with a sample of beneficiaries about whether the cream for babies should be zinc- or vaseline-based; the provision of spectacles, wheelchairs and walking sticks to elderly people, based on individual needs (even though this fell outside the entitlements accorded by the government's health policy); and recognition that for many elderly people, family contact was a higher priority than relief goods. Consequently all Red Cross Red Crescent staff were trained in tracing family members, and tracing and message forms were included in relief packages.

Other feedback, however, related to the desire of people to maintain their dignity. For example, they did not like the fact that: soup-kitchen areas were very visibly marked with the Red Cross Red Crescent emblem; baby clothes all came in one colour; and bulk food distribution was heavily marked with donor logos.

All of this, they pointed out, marked them as "refugees" and "victims" dependent on welfare handouts. The International Federation made changes accordingly. Their ability to respond to people's feedback was greatly enhanced by the presence of social workers on the programme staff, the active encouragement of field staff to seek out beneficiary feedback and adjust the programme accordingly, and flexibility from donors to amend programme design and contents of the relief package. ■

contextual and operational framework of reference to guide and assess humanitarian activities.

Secondly, this multiplication of interagency codes over the past decade followed a longer period during which individual agencies were less systematic in terms of organizational development. So once these standards and codes are consistently monitored, their validity and usefulness will be put to trial. Only those that emerge as relevant will become adopted and revised.

However, despite this multiplication of codes, the real risk to the quality and legitimacy of humanitarian work is currently posed by the absence of monitoring, and the lack of indicators by which to measure performance and compliance. Systematic monitoring will also allow for a critical reflection on the validity of different indicators in different situations, and therefore serve the purpose of learning.

## Box 7.7 Recommendations for the "accountable organization"

■ **Commit to human rights.** State a commitment to the protection and fulfilment of human rights. Provide adequate budgetary and human resources to realize this commitment.

■ **Set standards and indicators.** Set standards and performance indicators for protecting and fulfilling the rights of crisis-affected people and field staff. Set these in participation with stakeholders and review periodically.

■ **Communicate with all stakeholders.** Inform crisis-affected people and other stakeholders about standards adopted, aid programmes to be undertaken, and complaints processes available. Provide appropriate training in the use of standards.

■ **Involve crisis-affected people in programme management.** Involve affected people in the

planning, management and monitoring of aid programmes. Report to them regularly on the progress of programmes.

■ **Monitor compliance with standards.** Involve crisis-affected people and field staff in monitoring compliance with standards, and in revising them. Regularly audit compliance, using internal and external mechanisms.

■ **Resolve complaints.** Put in place complaints mechanisms, which safely and impartially provide crisis-affected people and field staff the opportunity to report concerns and to seek appropriate redress.

■ **Report back on standards.** Report back regularly to affected people and other stakeholders on compliance with standards and changes to programmes. ■

Systematic development of indicators and transparent monitoring of the implementation of standards and codes constitute the next, crucial step for humanitarian actors. And the results of monitoring must be made public and lead to tangible outcomes. Without all that, codes and standards will remain no more than paper tigers.

## Accountability mechanisms at field level

*"International actors must make sure that displaced communities are given a say in decisions that affect them. Displaced communities are not passive. They create their own strategies for addressing their needs by exchanging limited resources, services, information and shelter. Their involvement in identifying needs, in other decisions that affect their lives, and in implementing aid programmes is therefore essential."*

So argued the UN secretary-general to the Security Council in a report on the protection of civilians in armed conflicts in March 2001. That people displaced by disaster have their own coping strategies is hardly new. But recent experience in central

Africa and the Balkans makes the need to involve such communities at the planning stage essential, if aid providers are not to be left standing.

In 1999, for example, the spontaneous return into Kosovo of hundreds of thousands of refugees "wrongfooted" the cohorts of international relief agencies which had set up camps outside the territory. And more recently, following the volcanic eruption in the Democratic Republic of the Congo in January 2002, which destroyed most of the town of Goma, homeless people turned their backs on refugee camps constructed by international agencies in neighbouring Rwanda, preferring to brave the lava flows and salvage what they could from their old homes as quickly as possible.

Agencies' failure to engage crisis-affected people in meaningful dialogue about their needs and capacities can prove frustrating and even dangerous. One Luo chief was quoted in a 1996 article by Jok Madut Jok, entitled *Information Exchange in the Disaster Zone: Interaction between Aid workers and recipients in South Sudan*, as saying:

*"They came here several months ago and called us together, asked us a lot of questions that we did not know how to respond to. Then two months ago, one of these young girls came back with five other people and asked us the same questions for which we gave different responses each time. And now they are here again asking if Luo people have acquired cattle from the Dinka… We do not know what to tell them other than yes, some people have bought cows, others have not, and yes our people are hungry and you ought to help them. Bring food, more food."*

According to a paper published by the Humanitarian Practice Network in November 2000 evaluating NGO responses to Hurricane Mitch:

*"Decades of experiences notwithstanding, agencies still find it difficult to assess beneficiary needs adequately, and do not take sufficient note of local capacities and resources. Local participation in needs assessment, as well as in specific projects, must be strengthened if agencies are to provide aid that meets the needs of beneficiaries. Similarly, agencies should listen more carefully to the people they are trying to help when designing programmes."*

Many consultation exercises are more an exercise in extracting information than promoting dialogue. The listening aims at gathering some key data or filling information gaps about a population and the design of programmes. The listening is fairly superficial; information that would seriously question predetermined notions of programmes is likely to be censored. There often appears to be limited interest in the socio-economic and political dynamics of the affected or target populations, in the cultural and historical context. And there may be unwillingness to hand over decision-making power to affected communities or to question inappropriate codes and standards.

While many operational aid agencies may be experienced in listening to local people, there are some key questions that need to be asked. Listening and consultation for what? Who asks the questions? What happens next? Working in ways more transparent and accountable to crisis-affected people means more than just listening.

## Concerns and complaints

Complaints can be frivolous, or even fabricated for ulterior motives. But they can also be valid and serious (see Box 7.2). Most experienced field workers have heard one or several of these, often more than once. Yet they are rarely formally taken up within individual aid organizations or government aid administrations. The persistence of such concerns and complaints raises difficult questions. How can aid agencies better hear what people are concerned about, and distinguish between valid and invalid complaints? How can they tap the knowledge, experience and creativity of those affected by disaster, and draw from them positive ideas and suggestions for better programming? And respond to what they hear? What is under agencies' control, what can or can't they improve? How can they feed back to affected people what they can or can't do, what they plan to do and why they did what they did? How can they make this part of normal practice? And who is responsible for it? How can they hand over responsibility to "communities" if they cannot identify which actors represent "uncivil society"?

Many of these questions relate to the necessity to take into account the political and power relations between agencies, governments and crisis-affected populations and individuals, and to better unpack the various meanings, process and outcome of participation. So what can be done?

There are already examples of innovative approaches to foster greater transparency, responsiveness and accountability in programmes and policies. However, three specific components to accountability are beyond doubt: the obligations to *inform*, to *listen* and to *respond and report back* to crisis-affected populations.

## Obligation to inform

Agencies must inform affected people about aid operations, including issues such as agencies' mandates; the responsibilities they have willingly assumed; what is (or is not) under their authority; their programmes, budget situation, timetables, criteria, procedures, programme changes and evaluation outcomes; and agencies' policies and codes, including, for example, those with regard to sexual harassment.

Agencies must also inform beneficiaries about the relief they are entitled to and their rights.

Aid organizations can do this through: public meetings, leaflets or public notices, local radio, agency or interagency information centres, and information officers. Following the devastating earthquake in India's Gujarat state in January 2001, one grouping of local organizations took the rare step of stopping their material assistance programmes to concentrate on the provision of information, particularly about government policies, entitlements and procedures. They even provided a "help-desk" for people who had a claim or complaint for the civil service (see Box 7.3).

## Obligation to listen

Agencies must try to understand more about the population they are assisting and protecting, for example, their social and skills profiles, their history, knowledge, attitudes and practices. Agencies must actively seek the views of affected people on their perceived priorities and preferences; their concerns and complaints; what impacts they perceive aid programmes to have; how they see the relationship between aid-provider and aid-recipient; and constructive suggestions they might have to improve aid delivery or to help themselves better.

Humanitarian organizations need to understand how people themselves have struggled to cope with the crisis and how outside efforts can complement rather than undermine pre-existing coping strategies. They can do this through individual interviews, focus group discussions, participatory exercises, public meetings, sample surveys, opinion polls, a comments postbox or a complaints office or help-desk. Sometimes aid workers can simply discover things through observation.

Agencies can conduct participatory reviews or impact assessments with affected people. They can even commit to a regular "social audit" process with target groups or conduct population-wide social audits (see Box 7.4). They can suggest that evaluators are asked, in their terms of reference, to seek feedback from affected people, whether inside or outside target groups – and are given time to do so.

Crucially, aid-providers must set up complaints mechanisms, which safely and impartially provide individuals the opportunity to report concerns, make complaints and seek appropriate redress. In the northern Caucasus, the Danish Refugee Council has implemented a registration and complaints system for those affected by the conflict. Unusually the registration system, created on the basis of surveys, remains "alive" and up-to-date, allowing people to query and question the data (see Box 7.5).

chapter 7

## Obligations to respond and report back

What agencies have learned from those they try to assist and protect should stimulate many changes in the nature, design or implementation of aid programmes; the profile of aid staff; entitlement criteria; protection strategies; camp design; and even the nature of dialogue with affected people. It could provoke a search for better collaboration with other aid providers and duty-bearers, or lead agencies to lobby and support others as "amplifiers" of the voices which have been heard.

Aid-providers must assume responsibility, in a meaningful way, for what they did well, but also for where they failed, if it was under their control to do better. In order to demonstrate that listening to beneficiaries has resulted in concrete changes to aid provision, agencies must report back on their actions (and inactions) to crisis-affected communities, donors and other stakeholders. Reporting back completes the accountability circle of informing, listening and responding. Agencies can do this through sharing programme information and lessons with internal and external stakeholders, public campaigns, social audit reports, stakeholder surveys and assemblies, or public disclosures (e.g., web sites).

Following consultations with beneficiaries in the Balkans, the International Federation changed the style of its relief in a way which increased respect for the dignity of those affected by conflict and disaster (see Box 7.6).

The boxes on Sierra Leone (Box 7.1) and Nicaragua (Box 7.4) illustrate the initiatives in which an agency or a grouping of civil society organizations build an explicit link between the provision of aid and the wider politics of "good governance". The Nicaragua case illustrates how this can become controversial.

## Towards a culture of accountability

The challenge now is to turn still rather exceptional examples of accountability into regular, institutionalized practice. This is a challenge that requires integrating all three (field, headquarters and interagency) levels of decision-making and action. Field initiatives illustrated in this chapter should be supported through headquarters and interagency mechanisms, and funding policies committed to strengthening accountability. Recommendations for the "accountable humanitarian organization" are given in Box 7.7.

Many steps towards establishing a culture and practice of humanitarian accountability have already been made. Over the last year, debates and initiatives have taken place within the humanitarian community aimed at identifying better ways to monitor quality and accountability through, for instance, more rigorous self-assessment, peer review or accreditation. The way forward consists in building on these and older initiatives.

As human rights activists and ombudspersons have experienced, building a culture of accountability never ends; it is not a specific and tangible outcome but rather an ongoing process with a number of required benchmarks measuring its evolution. The process is very much driven by the articulation and acknowledgment of responsibilities or duties, and by the creation of specific mechanisms.

The agency and interagency standards and guidelines identified earlier offer a general framework within which humanitarian actors can define and limit their responsibilities. However, as Oxfam's Nicholas Stockton pointed out to a workshop in 1996, "Occupational performance standards do not on their own ensure quality control... The demand for rapid recruitment in emergencies does result in inexperienced and unsuitable staff being pitched into the most demanding and responsible occupational contexts, with negligible specialist induction and support."

Furthermore, such standards, codes and benchmarks are not disseminated as widely as they should be. And very few of them are evaluated, monitored or their non-implementation sanctioned. The absence of meaningful self-regulation at every level diminishes the impact that these initiatives could have, in terms of offering an updated and context-sensitive interpretation of humanitarian responsibilities.

Is it even sufficient to rely solely on self-regulatory mechanisms? The majority of professional sectors (medicine, judiciary, police, public administration), which have adopted a self-regulatory framework, also recognize that a system without external and independent regulatory, monitoring and response mechanisms may be neither principled nor effective.

Similarly, however, relying solely on independent mechanisms will not be sufficient or effective either. Among other things, it would give sole regulatory authority to mechanisms or bodies, whose modus operandi may differ greatly from that of the humanitarian sector. Independent mechanisms or bodies should be called upon when the issues to be considered are beyond the mandate of self-regulatory bodies, or as part of an appeal process.

Strengthening accountability is both an individual agency responsibility and a collective responsibility, shared by all civilian humanitarian actors. The sector, as a whole, must acknowledge, through self-regulatory and independent monitoring bodies and mechanisms, its responsibility to ensure that its members observe minimum standards in humanitarian response. In particular, humanitarian actors should act together to build self-regulatory bodies, at national and international levels, which should:

- ensure beneficiary participation;
- ensure other stakeholders' participation;

- be transparent;
- have the mandate or authority to monitor and enforce agreed rules;
- update rules;
- accredit or remove the endorsement of non-complying organizations; and
- include a right of appeal.

While regulation is clearly crucial, it is equally important that agencies and their employees are geared towards feeling and being accountable to crisis-affected populations. It is imperative that affected people, on whose behalf humanitarian action is conducted and funds raised, are able to exercise their rights to information, to have a say in decisions that affect them, and to seek redress where appropriate.

Only by being transparent in its undertakings and accountable to those whose lives it most affects, can humanitarian action truly meet its objective to safeguard and uphold the well-being and dignity of those who have been affected by disasters and armed conflicts.

*Agnès Callamard and Koenraad Van Brabant, co-directors of the Humanitarian Accountability Project, were principal contributors to this Chapter and all boxes except Box 7.1, which was contributed by Steve Archibald (CARE International) and Paul Richards, University of Wageningen (Netherlands).*

## Sources and further information

Amnesty International. *Annual Report* 1998, London: Amnesty International, 1998.

Amnesty International. *Broken Bodies, Shattered Minds: Torture and Ill-Treatment of Women.* London: Amnesty International, 2001.

Borton, John. *The Quality revolution and some lessons on what the humanitarian sector might learn from it*, a paper presented at a workshop on "Quality in Humanitarian Aid", Göttingen, 28-30 September, 2001.

Committee on Economic, Social and Cultural Rights. *The rights to adequate food (Art.11), General Comment 12*, E/C.12/1999/5, CESCR, 12 May 1999.

Cuny, Fred. *Disasters and Development*, 1983.

Goma Epidemiology Group. "Public health impact of Rwandan refugee crisis: what happened in Goma, Zaire, in July 1994?" in *The Lancet.* 1995.

Grunewald, F., de Geoffroy, V. and Lister, S. *NGO Responses to Hurricane Mitch: Evaluations for Accountability and Learning,* a Humanitarian Practice Network paper, ODI, November 2000.

Humanitarian Accountability Project (HAP). *Why Humanitarian Accountability? Contextual and Operational Factors.* Geneva: HAP, 2001.

Humanitarian Accountability Project. *Recommendations for an accountable organisation.* Briefing 6, April 2002.

International Council on Human Rights Policy (ICHRP). *Taking Duties Seriously: Individual Duties in International Human Rights Law*, Geneva: ICHRP, 1999, pp.13-18.

International Federation of Red Cross and Red Crescent Societies. *Code of Conduct for the International Red Cross and Red Crescent Movement and NGOs in Disaster Relief*. Geneva: International Federation, 1994. http://www.ifrc.org/publicat/conduct/

Joint Evaluation of Emergency Assistance to Rwanda. *Humanitarian Aid and Effects*. March 1996.

Jok, Jok Madut. "Information Exchange in the Disaster Zone: Interaction between Aid workers and recipients in South Sudan", in *Disasters*, Vol. 20, No. 3., 1996.

O'Brien, Paul. "Promoting Rights and Responsibilities" in *CARE Newsletter*, February 2001.

Raynard, Peter. *Mapping Accountability in humanitarian assistance*. Active Learning Network for Accountability and Practice (ALNAP), May 2000.

Salama, P., Buzard, N. and Spiegel, P. "Improving standards in international humanitarian response: the Sphere project and beyond" in *Journal of the American Medical Association*, Vol. 285, No. 85, August 1, 2001.

Slim, Hugo. "Not Philanthropy But Rights" in *International Journal of Human Rights,* Vol. 6, 2002.

Stockton, Nicholas. *"Humanitarian Standards – Rations or Rights",* a paper presented at the International Oxfam Emergency Workshop, Brussels, May 1996.

UN High Commissioner for Human Rights (UNHCHR), *Business and Human Rights: A Progress Report*. Geneva: UNHCHR, 2000. http://www.unhchr.ch/business.htm

UN Office for the Coordination of Humanitarian Affairs (OCHA). *Protection of Civilians in Armed Conflicts, Protection Parameters.* 1999.

United Nations. *Report of the UNSG to the Security Council on the protection of civilians in armed conflicts.* S/2001/331, March 2001.

Yaansah, Eddie Adlin. "Code of Conduct for NGOs in Ethiopia" in *The International Journal of Not-for-profit Law,* Vol. 1, no. 3, 1999.

## Web sites

ALNAP http://www.alnap.org
CARE http://www.care-international.org
ReliefWeb (OCHA) http://www.reliefweb.int
UNHCHR http://www.unhchr.ch

chapter 8

Section Two
**Tracking
the system**

# Disaster data: key trends and statistics

This year's *World Disasters Report* features the latest verified data on both "natural" and technological disasters from the past decade (1992-2001). Our natural disaster data are divided into hydro-meteorological and geophysical disasters, in order to track the changing impact of weather-related hazards. And we analyse the data not only by country and continent, but according to levels of human development, to highlight the relationship between development and disasters.

## Weather-related disasters continue to soar

While the total number of all disasters reported during 2001 was lower than the previous year, at 712 events it still represents the second highest total of the decade (see Table 1). The overall figure disguises some disturbing trends. Countries of low human development (LHD) reported an increase in disasters. Africa reported twice as many disasters last year compared to the decade's average of 94 events.

From 1992-1996 an average of 75 floods per year was reported. But in 2001, for the second year running, the number of reported floods was more than double that figure (see Table 5). While the number of geophysical disasters has remained fairly constant, the past two years have seen the highest numbers of weather-related disasters reported over the decade. However, while half of all disasters reported in 2001 were hydro-meteorological, nearly one-third were transport accidents. In Africa, 46 per cent of all reported disasters during the decade were transport related (see Table 9).

## Deaths double between 2000 and 2001

A total of 39,073 people were reported killed by disasters in 2001 (see Table 2). While this was nearly double the figure for the previous year, it was lower than the decade's annual average of around 62,000. Most of the fatalities for 2001 were reported in countries of medium human development (MHD), which registered an increase in deaths from disasters of 25 per cent compared to the ten-year average.

Last year, earthquakes proved to be the world's deadliest disasters, accounting for over half the year's toll (see Table 6). Much of this can be attributed to the quakes which hit the Indian state of Gujarat in January 2001. For the first year in the decade, geophysical disasters killed more people than weather-related disasters. Over the decade, however, hydro-meteorological hazards have claimed 71 per cent of all lives lost to disasters (see Figure 8.1).

Photo opposite page: Behind the data are individual men, women and children, each with a story to tell, a life to lead or a livelihood to lose. Getting the numbers right helps improve the efficiency and effectiveness with which humanitarian agencies serve vulnerable people.

Yoshi Shimizu/ International Federation, Sierra Leone 2001.

Figure 8.1
Source: CRED.

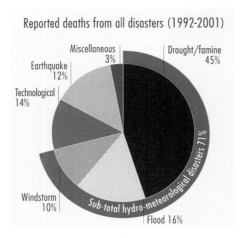

Reported deaths from all disasters (1992-2001)

Miscellaneous 3%
Earthquake 12%
Technological 14%
Drought/famine 45%
Windstorm 10%
Flood 16%
Sub-total hydro-meteorological disasters 71%

From 1992-2001, LHD countries have accounted for just one-fifth of the total number of disasters, but over half of all disaster fatalities. On average 13 times more people die per reported disaster in LHD countries than in countries of high human development (HHD). Over the decade, different kinds of disaster have proved deadly in different continents (see Table 10). In the Americas, floods accounted for 45 per cent of all deaths from disasters. In Asia, drought/famine claimed 58 per cent. In Europe, earthquakes claimed 58 per cent, while in Oceania, tidal waves claimed 66 per cent. Surprisingly, Africa's deadliest disasters were transport accidents – claiming 45 per cent of the decade's deaths.

## Disasters affect 170 million worldwide

Last year, a total of 170 million people were reported affected by disasters – below the decade's average of 200 million (see Table 3). However, this disguises a figure of 11 million affected in the Americas in 2001, more than double that continent's annual average.

Earthquakes affected more people during 2001, 19 million, than any other year of the decade (see Table 7). Meanwhile drought/famine affected over 86 million people last year, many of those living in central and south Asia. Over the decade, however, floods accounted for nearly two-thirds of all those affected by disasters (see Figure 8.2).

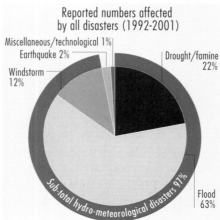

Reported numbers affected by all disasters (1992-2001)

Miscellaneous/technological 1%
Earthquake 2%
Windstorm 12%
Drought/famine 22%
Flood 63%
Sub-total hydro-meteorological disasters 97%

Weather-related disasters have been the most pervasive in the past ten years. Drought/famine accounted for 82 per cent of all those affected in Africa, 48 per cent in Oceania and 35 per cent in the Americas. Meanwhile, floods accounted for 69 per cent of all those affected in Asia. And windstorms accounted for 36 per cent of those affected in the Americas, and 33 per cent in Europe (see Table 11).

Figure 8.2
Source: CRED.

# Disaster damage below decade's average

The total amount of estimated damage (direct damage to infrastructure, crops, etc.) inflicted by disasters during 2001 was US$ 24 billion – the decade's lowest and well below the annual average of US$ 69 billion (see Table 4). Two-thirds of 2001's damage was reported from MHD countries. Earthquakes caused an estimated US$ 9 billion of damage and windstorms US$ 8 billion last year (see Table 8).

Over the decade, earthquakes have proved the most expensive of disasters, costing the world US$ 238 billion in damage alone – without even measuring the effect on economies. Around half of this figure, however, can be attributed to one event – the 1995 quake in Kobe, Japan. Globally, floods and windstorms are very nearly as costly as earthquakes (see Figure 8.3).

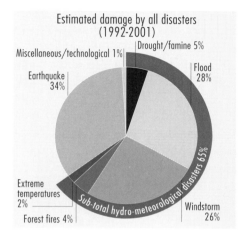

Estimated damage by all disasters (1992-2001)

Disasters can have a devastating effect on the development of poorer nations. In Honduras, for example, Hurricane Mitch put the country's economic development back 20 years. In 1998, an El Niño year, Peru suffered storm damage to public infrastructure estimated at equivalent to 5 per cent of gross domestic product (GDP). In 1999, losses from earthquakes in Turkey and landslides in Venezuela cost these countries the equivalent to 10 per cent of their GDP.

Looking at the different continents, the costliest disasters over the past ten years have been floods in Africa (35 per cent) and Europe (37 per cent), windstorms in the Americas (54 per cent), earthquakes in Asia (46 per cent) and drought/famine in Oceania (47 per cent) (see Table 12).

**Figure 8.3**
Source: CRED.

# Disasters kill fewer but affect more

Over the past 30 years, the impacts of natural disasters have changed dramatically (see Chapter 1, Figure 1.1). Deaths from natural disasters fell from nearly 2 million in the 1970s to just under 800,000 in the 1990s. But numbers reported affected by natural disasters rocketed from just over 700 million in the 1970s to nearly 2 billion in the 1990s.

The reasons behind these statistics are complex and need further analysis. However, the drop in fatalities can be attributed in part to better disaster preparedness. In 1970 a cataclysmic cyclone killed half a million people in Bangladesh – accounting for a

## Box 8.1 EM-DAT: a specialized disaster database

The Centre for Research on the Epidemiology of Disasters (CRED), established in 1973 as a non-profit institution, is located at the School of Public Health of the Louvain Catholic University in Brussels, Belgium. CRED became a World Health Organization (WHO) collaborating centre in 1980. Although CRED's main focus is on public health, the centre also studies the socio-economic and long-term effects of large-scale disasters.

Since 1988, CRED has maintained an Emergency Events Database (EM-DAT), sponsored by the International Federation, WHO, the United Nations Office for the Coordination of Humanitarian Affairs (OCHA) and the European Community Humanitarian Office (ECHO). USAID's Office of Foreign Disaster Assistance (OFDA) also collaborated in getting the database started, and a recent OFDA/CRED initiative has made a specialized, validated disaster database available on CRED's web site. The database's main objective is to assist humanitarian action at both national and international levels and aims at rationalizing decision-making for disaster preparedness, as well as providing an objective base for vulnerability assessment and priority setting.

Tables 1 to 13 in this chapter have been drawn from EM-DAT, which contains essential core data on the occurrence and effects of over 12,000 disasters in the world from 1900 to the present. The database is compiled from various sources, including UN agencies, NGOs, insurance companies, research institutes and press agencies. The entries are constantly reviewed for redundancies, inconsistencies and the completion of missing data.

CRED consolidates and updates data on a daily basis; a further check is made at three-monthly intervals; and revisions are made annually at the end of the calendar year. Priority is given to data from UN agencies, followed by OFDA, and then governments and the International Federation. This priority is not a reflection on the quality or value of the data, but the recognition that most reporting sources do not cover all disasters or have political limitations that may affect the figures. ∎

Dr. D. Guha-Sapir
Director, Centre for Research on the Epidemiology of Disasters (CRED)
WHO Collaborating Centre
School of Public Health
Catholic University of Louvain
30.94, Clos Chapelle-aux-Champs
1200 Brussels
Belgium
Tel.: 32 2 764 3327
Fax: 32 2 764 3441
E-mail: sapir@epid.ucl.ac.be
Web: http://www.cred.be

quarter of that decade's fatalities from natural disasters. Following that catastrophe, the Bangladesh government supported by the Red Cross Red Crescent initiated the cyclone preparedness programme (CPP) – a system of early warning and evacuation that has proved enormously successful (see Chapter 1, Box 1.1). In the 1990s alone, the CPP successfully evacuated 2.5 million people into emergency shelters before cyclones hit – and very probably saved their lives as a result.

chapter 8

Many factors are likely to be contributing to the increase in those reported affected by disasters. The profile of vulnerability is changing. As more people move into urban areas and slum settlements, they are increasingly living in the path of disaster. Traditional coping mechanisms are being eroded as families fragment and communities disperse. Environmental degradation is increasing the negative effects of floods, windstorms and droughts. While disaster preparedness measures are helping save lives, the failure to reduce risks more broadly may be contributing to the higher numbers of disaster-affected people. Better reporting of the numbers of disaster-affected people may contribute to the higher figure. And as pointed out below, the definition of "affected" is open to interpretation.

This global trend of lower deaths and higher numbers of affected is mirrored by the past two decades of data on all disasters (natural and technological). Worldwide, such disasters claimed 1 million lives from 1982-91. This total fell by 40 per cent to around 620,000 deaths from 1992-2001 (see Table 13). But the numbers affected by all disasters climbed from 1.7 billion (1982-91) to 2 billion (1992-2001).

However, these global figures disguise some serious discrepancies between the continents. The fall in global disaster fatalities over the past two decades is largely due to an enormous drop in African deaths, from 575,000 (1982-91) to 40,000 (1992-2001). The high fatalities from the earlier decade relate to the series of famines which devastated the Horn of Africa in the mid-1980s. Apart from Africa and Europe, the rest of the world reported substantial increases in the numbers of disaster fatalities in the past two decades. For Oceania, deaths tripled from one decade to the next, while for Asia deaths were up 41 per cent and for the Americas up 32 per cent. Meanwhile, the figures for those affected have more than tripled in Europe and increased 12-fold in Oceania.

## Development and emergency aid fall

Official development assistance (ODA) from members of the Organisation for Economic Co-operation and Development's (OECD) Development Assistance Committee (DAC) slipped to US$ 53.7 billion during 2000, the latest year for which statistics are available (see Figure 8.4). This represented a drop of around US$ 2.7 billion from aid disbursements during 1999. Some of this decrease was accounted for by Japan, the world's largest donor, which slashed its aid by US$ 1.8 billion. France cut its aid by over one-quarter to the lowest level of the past decade. Nine out of the 22 DAC donors gave more than in 1999, notably the United Kingdom, which increased its aid by over US$ 1 billion, a 30 per cent increase.

Expressed as a percentage of donor countries' gross national product (GNP), ODA remained static during 2000 at 0.39 per cent (see Figure 8.5). Denmark, the Netherlands, Norway and Sweden remained the most generous donors, exceeding the

Figure 8.4
Source:
CRED/OECD.

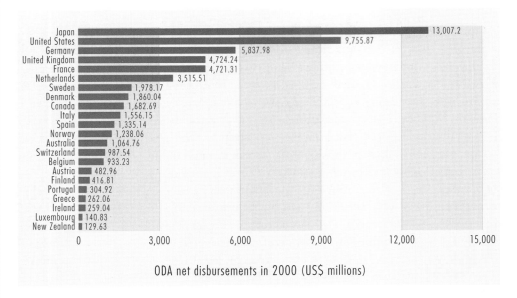

ODA net disbursements in 2000 (US$ millions)

United Nations (UN) target of 0.7 per cent, as they have done for the past decade. But 2000 saw a new member of the exclusive "0.7 per cent club": Luxembourg, which steadily increased its development assistance from 0.21 per cent of GNP in 1990 to 0.71 per cent in 2000. The United States stayed at the bottom of the pile, donating one-tenth of 1 per cent of its US$ 10 trillion GNP in aid.

Emergency/distress relief from DAC donors fell from the decade's high of US$ 4.4 billion in 1999 to US$ 3.6 billion in 2000 (see Figure 8.6). However, 2000's figure was

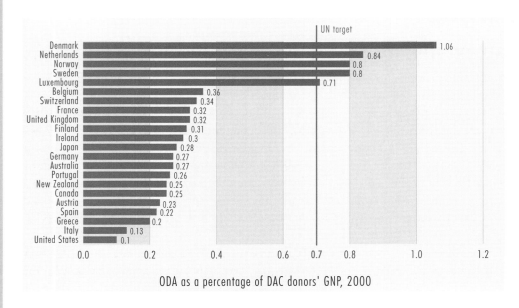

ODA as a percentage of DAC donors' GNP, 2000

Figure 8.5
Source:
CRED/OECD.

still the second highest of the decade. The biggest donor was the United States, which accounted for one-third of all emergency aid donations.

## Sea-change in attitudes to aid?

Following the attacks in the United States on 11 September 2001, a number of world leaders suggested that the fight against poverty could help create a more secure global environment. James Wolfensohn, president of the World Bank, was one of the first to make the connection between terrorism and poverty. Just weeks after the attacks, he said, "We estimate that tens of thousands more children will die worldwide and about 10 million more people are likely to be living below the poverty line of $1 a day because of the terrorist attacks."

In February 2002, Colin Powell, US secretary of state, argued before a meeting of business leaders that "terrorism really flourishes in areas of poverty, despair and hopelessness, where people see no future". Wolfensohn added that "hard-headed politicians" should act out of self-interest and view greater financial aid to poor countries "as an insurance policy against future terrorism".

At the same time, the UN's secretary-general, Kofi Annan, called on governments to raise an extra US$ 50 billion a year in ODA to improve the chances of attaining the international development goals of 2015. These include halving the numbers of people living in hunger and poverty, ensuring that all children complete primary education, and halting the spread of AIDS. Annan said that this increase in rich-country aid, which would amount to a doubling of ODA, was "an immediate, short-term tar-

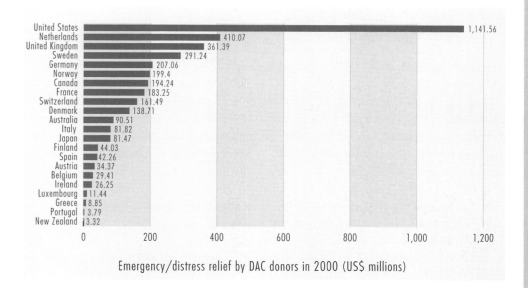

Emergency/distress relief by DAC donors in 2000 (US$ millions)

**Figure 8.6**
Source:
CRED/OECD.

## Box 8.2 US Committee for Refugees

USCR is the public information and advocacy arm of Immigration and Refugee Services of America, a non-governmental organization. USCR's activities are twofold: it reports on issues affecting refugees, asylum seekers, and internally displaced people; and encourages the public, policy-makers, and the international community to respond appropriately and effectively to the needs of uprooted populations.

USCR travels to the scene of refugee emergencies to gather testimony from uprooted people, to assess their needs, and to gauge governmental and international response. The committee conducts public briefings to present its findings and recommendations, testifies before the United States Congress, communicates concerns directly to governments, and provides first-hand assessments to the media. USCR publishes the annual World Refugee Survey, the monthly Refugee Reports and issue papers. ■

US Committee for Refugees
1717 Massachusetts Avenue, NW, Suite 200
Washington, DC 20036
USA
Tel: + 1 202 347 3507
Fax: + 1 202 347 3418
E-mail: uscr@irsa-uscr.org
Web: http://www.refugees.org

get, to be achieved within two or three years". Annan compared humanity to passengers on a small, storm-tossed boat: "If they are sick, we all risk infection, and if they are hungry, all of us can easily get hurt."

In March 2002, the World Bank echoed Annan's call for a doubling of world aid, and its president called on rich countries to lower their trade barriers to products from poor countries. Farming subsidies paid by wealthy countries are six times greater than foreign aid flows, he said. Some critics of the United States' government have pointed out that the average US cow is subsidized by US$ 600 per year – double the amount on which over 1 billion human beings subsist.

The rhetoric was ratcheted up in preparation for the Monterrey conference in March 2002, which saw 50 world leaders meet in Mexico to review the state of development finance. Mike Moore, head of the World Trade Organization, told delegates, "Poverty in all its forms is the greatest single threat to peace, security, democracy, human rights and the environment."

The Monterrey meeting failed to produce the doubling in aid pledges that the UN and World Bank had sought. But some of the world's biggest donors made moves in the right direction. The European Union agreed to boost its aid budgets by US$ 7 billion by 2006. And the US pledged to increase foreign aid spending by 50 per cent, or US$ 5 billion, in the three years from 2004.

However, the US administration has made it clear that they regard aid as contingent on political and economic reforms. Kofi Annan expanded on this new "global deal" in Mexico: "Where developing countries adopt market-oriented policies, strengthen their institutions, fight corruption, respect human rights and the rule of law, and spend more on the needs of the poor, rich countries can support them with trade, aid, investment and debt relief." And while the amounts pledged in Monterrey were not enough, Annan concluded, "These decisions do suggest that the argument on principle has now been won. All governments accept that official aid is only one element in the mix, but an essential one. Aid is much more effective than it was 20 years ago."

## Disaster data: handle with care

Data on disaster occurrence, its effect upon people and its cost to countries remain, at best, patchy. No single institution has taken on the role of prime providers of verified data, so the *World Disasters Report* draws upon two main sources: the Centre for Research on the Epidemiology of Disasters (CRED) and the US Committee for Refugees (USCR) (see Boxes 8.1 and 8.2). CRED has used the OECD's DAC database for ODA and emergency/distress relief statistics.

Key problems today with disaster data include the lack of standardized collection methodologies and definitions. Problems exist over such loose categories as "internally displaced" people or even people "affected" by disaster. Much of the data in this chapter, except that on DAC spending, is culled from a variety of public sources: newspapers, insurance reports, aid agencies, etc. The original information is not specifically gathered for statistical purposes and so, inevitably, even where the compiling organization applies strict definitions for disaster events and parameters, the original suppliers of the information may not. The figures therefore should be regarded as indicative. Relative changes and trends are more useful to look at than absolute, isolated figures.

Information systems have improved vastly in the last 25 years and statistical data is now more easily available. However, the lack of systematic and standardized data collection from disasters in the past is now revealing itself as a major weakness for any long-term planning. Despite efforts to verify and review data, the quality of disaster databases can only be as good as the reporting system. Fortunately, due to increased pressures for accountability from various sources, many donor and development agencies have started placing priority on data collection and its methodologies, but this has yet to result in any recognized and acceptable international system for disaster-data gathering, verification and storage.

Dates can be a source of ambiguity. For example, the declared date for a famine is both necessary and meaningless – famines do not occur on a single day. In such

## Box 8.3 GLobal IDEntifier Number (GLIDE)

Accessing disaster information can be a time-consuming and laborious task. Not only is data scattered but frequently identification of the disaster can be confusing in countries with many disaster events. To address these problems, CRED, together with their technical advisory group partners (ReliefWeb-OCHA, the Asian Disaster Reduction Center-Kobe, OFDA-USAID, the UN's Food and Agricultural Organization, the National Oceanic and Atmospheric Administration-Office of Global Programs and the World Bank) have launched a new initiative: GLIDE.

A GLobal IDEntifier number (GLIDE) is issued every week by EM-DAT at CRED whose net captures all the disasters that fulfil the EM-DAT criteria. The components of a GLIDE number consist of two letters to identify the disaster type (e.g., ST for storms); the year of the disaster; a four-digit, sequential disaster number; and the three-letter ISO code for country of occurrence. So, for example, the GLIDE number for Hurricane Mitch in Honduras is: ST-1998-0345-HND.

This number is posted by ReliefWeb on all their documents relating to that particular disaster and gradually all other partners will include it in whatever information they generate. As information suppliers join in this initiative, documents and data may be captured by any web search engine based on one single number with high specificity. The success of GLIDE depends on its widespread use and its level of utility for practitioners.

Today, users all over the world can pick up the GLIDE number from the home pages of CRED (http://www.cred.be) and ReliefWeb (http://www.reliefweb.int). At present in an experimental phase, the group welcomes comments or suggestions. For more information, please contact: sapir@epid.ucl.ac.be ∎

cases, the date the appropriate body declares an official emergency has been used. Changes in national boundaries also cause ambiguities in the data, most notably the break-up of the Soviet Union and Yugoslavia, and the unification of Germany.

Data can be skewed because of the rationale behind data gathering. Reinsurance companies, for instance, systematically gather data on disaster occurrence in order to assess insurance risk, but only in areas of the world where disaster insurance is widespread. Their data may therefore miss out poorer disaster-affected regions where insurance is unaffordable or unavailable.

Data on the numbers of people affected by a disaster can provide some of the most potentially useful figures, for planning both disaster preparedness and response, yet these are also some of the most loosely reported figures. The definition of "affected" is open to interpretation, political or otherwise. In conflict, warring parties may wish to maximize sympathy for their causes and exaggerate the numbers of people under their control who are said to be affected. Even if political manipulation is absent, data is often extrapolated from old census information, with assumptions being made about percentages of an area's population affected.

Part of the solution to this data problem lies in retrospective analysis. Data is most often publicly quoted and reported during a disaster event, but it is only long after the event, once the relief operation is over, that estimates of damage and death can be verified. Some data gatherers do this, and this accounts for retrospective annual disaster figures changing one, two and sometimes even three years after the event.

# Methodology and definitions

The *World Disasters Report* divides disasters into the following types:

## Natural disasters

- **Hydro-meteorological:** avalanches/landslides; droughts/famines; extreme temperatures; floods; forest/scrub fires; windstorms; and other (insect infestation and waves/surges).
- **Geophysical:** earthquakes; volcanic eruptions.

## Non-natural disasters

- **Industrial:** chemical spill, collapse of industrial structures, explosion, fire, gas leak, poisoning, radiation.
- **Miscellaneous:** collapse of domestic/non-industrial structures, explosion, fire.
- **Transport:** air, rail, road and water-borne accidents.

CRED uses the following definitions for disaster and conflict:

## Disaster

A situation or event, which overwhelms local capacity, necessitating a request to national or international level for external assistance. In order for a disaster to be entered in EM-DAT at least one of the following criteria has to be fulfilled:

- 10 or more people reported killed;
- 100 people reported affected;
- a call for international assistance; and/or
- declaration of a state of emergency.

**Killed:** People confirmed dead, or missing and presumed dead.

**Affected:** People requiring immediate assistance during a period of emergency, i.e., requiring basic survival needs such as food, water, shelter, sanitation and immediate

chapter 8

medical assistance. In EM-DAT, the total number of people affected include people reported injured, homeless, and affected.

**Estimated damage:** The economic impact of a disaster usually consists of direct damage (e.g., to infrastructure, crops, housing) and indirect damage (e.g., loss of revenues, unemployment, market destabilization). EM-DAT's estimates relate only to direct damage.

## Refugees, asylum seekers and internally displaced people

The data in Tables 14, 15 and 16 were provided by USCR, and concern three categories of uprooted people: refugees, asylum seekers and internally displaced people. Data concerning these populations are often controversial because they involve judgements about why people have left their home areas. Differing definitions of the groups in question often promote confusion about the meaning of reported estimates.

USCR does not conduct censuses of these populations, although it does conduct firsthand site visits to assess refugee conditions. The committee evaluates population estimates circulated by governments, UN agencies and humanitarian assistance organizations, and discerns which of the various estimates appear to be most reliable. The estimates reproduced in these tables are USCR's preliminary year-end figures for 2001.

The quality of the data in these tables is affected by the less-than-ideal conditions often associated with flight. Unsettled conditions, the biases of governments and opposition groups, and the need to use population estimates to plan for providing humanitarian assistance can each contribute to inaccurate estimates.

**Tables 14 and 15** concern refugees and asylum seekers: Table 14 lists refugees and asylum seekers by country of origin, while Table 15 lists the two groups by host country. Refugees are people who are outside their home country and are unable or unwilling to return to that country because they fear persecution or armed conflict. Asylum seekers are people who claim to be refugees; many are awaiting a determination of their refugee status. While not all asylum seekers are refugees, they are nonetheless entitled to certain protections under international refugee law, at least until they are determined not to be refugees.

Different standards for refugee status exist in different countries or regions. Recognition of refugee status, however, does not make someone a refugee, but rather declares her or him to be one. "He does not become a refugee because of recognition, but is recognized because he is a refugee," the UN High Commissioner for Refugees (UNHCR) has noted. Not all refugees are recognized as such by governments.

USCR includes in Table 15 people who have been admitted as refugees or granted asylum during the year, but thereafter regards them as having been granted permanent protection, even if they have not yet officially become citizens of their host country. This method of record-keeping differs from that employed by UNHCR, which continues counting refugees until they gain citizenship.

**Table 16** concerns internally displaced people. Like refugees and asylum seekers, internally displaced people have fled their homes; unlike refugees and asylum seekers, however, internally displaced people remain within their home country.

No universally accepted definition of an internally displaced person exists. USCR generally considers people who are uprooted within their country because of armed conflict or persecution – and thus would be refugees if they were to cross an international border – to be internally displaced. Broader definitions are employed by some agencies, however, who sometimes include people who are uprooted by natural or man-made disasters or other causes not directly related to human rights.

Internally displaced people often live in war-torn areas and may be subject to ongoing human rights abuse, sometimes at the hands of their own government. Most of the internally displaced are neither registered nor counted in any systematic way. Estimates of the size of internally displaced populations are frequently subject to great margins of error.

*In the following tables, some totals may not correspond due to rounding.*

## Erratum – World Disasters Report 2001

1. Chapter 8, Table 13. An error occurred during the compilation of data for Oceania. Reported disaster impacts were incorrect for 11 Oceania states: Kiribati, Marshall Islands, Federated States of Micronesia, New Caledonia, New Zealand, Niue, Palau, Papua New Guinea, Samoa, Solomon Islands and Tokelau. This has some implications for 2001's map: in particular, the Solomon Islands, which were reported as being the most disaster-affected country, are no longer top of the table. Malawi therefore emerges as the most disaster-affected nation from 1991-2000. For a corrected version of Table 13, please contact CRED. For a corrected version of the map table, please contact the International Federation.

2. Chapter 8, Table 13. Two other data discrepancies were noticed in the 1981-1990 decade. In Africa, Sao Tome &

Principe and Senegal data were wrongly attributed respectively to Senegal and Seychelles. In Asia, Viet Nam and Yemen data were wrongly attributed respectively to Uzbekistan and Viet Nam. For a corrected version of these tables, please contact CRED.

3. Chapter 8, Tables 3, 7, 11, 13. An error occurred during the compilation of data for windstorms in the United States during 1999. The total number of people affected was overreported by 8,738,000. This number must therefore be subtracted from the totals of people affected as reported in Tables 3, 7, 11, 13. For corrected versions of these tables, please contact CRED.

CRED would like to apologize for any misunderstanding caused by the misreporting of these data. All data for tables in the *World Disasters Report 2002* have been adjusted to take account of these discrepancies. ∎

**Table 1** Total number of reported disasters, by continent and by year (1992 to 2001)

| | 1992 | 1993 | 1994 | 1995 | 1996 | 1997 | 1998 | 1999 | 2000 | 2001 | Total |
|---|---|---|---|---|---|---|---|---|---|---|---|
| Africa | 55 | 49 | 58 | 61 | 58 | 56 | 84 | 144 | 195 | 186 | 946 |
| Americas | 88 | 94 | 81 | 97 | 93 | 100 | 112 | 135 | 144 | 127 | 1,071 |
| Asia | 162 | 221 | 181 | 171 | 176 | 194 | 206 | 239 | 286 | 292 | 2,128 |
| Europe | 51 | 46 | 69 | 64 | 54 | 60 | 65 | 78 | 121 | 90 | 698 |
| Oceania | 12 | 14 | 17 | 8 | 17 | 15 | 18 | 15 | 13 | 17 | 146 |
| High human development | 99 | 112 | 93 | 97 | 90 | 112 | 106 | 111 | 155 | 135 | 1,110 |
| Medium human development | 202 | 245 | 254 | 232 | 229 | 238 | 277 | 363 | 430 | 396 | 2,866 |
| Low human development | 67 | 67 | 59 | 72 | 79 | 75 | 102 | 137 | 174 | 181 | 1,013 |
| **Total** | **368** | **424** | **406** | **401** | **398** | **425** | **485** | **611** | **759** | **712** | **4,989** |

Source: EM-DAT, CRED, University of Louvain, Belgium

Asia remains the continent most frequently struck by disaster, with 42.6 per cent of the total number of events recorded in EM-DAT. After a steady increase from 1992 to 2000, the number of disaster events fell to 712 in 2001. Countries classed as medium human development accounted for more than half of the decade's disasters.

chapter 8

**Table 2** Total number of people reported killed by disasters, by continent and by year (1992 to 2001)

| | 1992 | 1993 | 1994 | 1995 | 1996 | 1997 | 1998 | 1999 | 2000 | 2001 | Total |
|---|---|---|---|---|---|---|---|---|---|---|---|
| Africa | 4,981 | 1,637 | 3,144 | 2,962 | 3,484 | 4,003 | 7,092 | 2,688 | 5,610 | 4,475 | 40,076 |
| Americas | 1,748 | 4,606 | 2,925 | 2,622 | 2,530 | 2,753 | 22,944 | 33,957 | 1,759 | 3,449 | 79,293 |
| Asia | 13,414 | 22,787 | 13,361 | 74,975 | 69,704 | 71,091 | 82,391 | 75,890 | 11,087 | 28,981 | 463,681 |
| Europe | 2,115 | 1,162 | 2,340 | 3,366 | 1,204 | 1,166 | 1,429 | 19,448 | 1,605 | 2,159 | 35,994 |
| Oceania | 6 | 120 | 103 | 24 | 111 | 398 | 2,227 | 116 | 205 | 9 | 3,319 |
| High human development | 918 | 1,935 | 2,486 | 7,825 | 2,009 | 1,800 | 2,167 | 4,447 | 1,814 | 2,063 | 27,464 |
| Medium human development | 13,907 | 22,805 | 15,470 | 17,330 | 14,982 | 17,779 | 43,618 | 70,302 | 12,305 | 32,635 | 261,133 |
| Low human development | 7,439 | 5,572 | 3,917 | 58,794 | 60,042 | 59,832 | 70,298 | 57,350 | 6,147 | 4,375 | 333,766 |
| **Total** | **22,264** | **30,312** | **21,873** | **83,949** | **77,033** | **79,411** | **116,083** | **132,099** | **20,266** | **39,073** | **622,363** |

Source: EM-DAT, CRED, University of Louvain, Belgium

Of the total number of people reported killed by disasters, 74.5 per cent are in Asia. More than half of the fatalities occurred in nations of low human development, with high human development countries accounting for only 4.4 per cent of deaths. In Asia in 2001, the Gujarat earthquake killed some 20,000 people, almost 70 per cent of the continent's total deaths. The worst year for fatalities was 1999, when 30,000 people were killed in the Venezuelan floods and 17,000 died in earthquakes in Turkey.

**Table 3** Total number of people reported affected by disasters, by continent and by year (1992 to 2001) in thousands

| | 1992 | 1993 | 1994 | 1995 | 1996 | 1997 | 1998 | 1999 | 2000 | 2001 | Total |
|---|---|---|---|---|---|---|---|---|---|---|---|
| Africa | 21,747 | 16,500 | 11,234 | 9,533 | 4,282 | 7,136 | 10,738 | 14,362 | 23,043 | 18,378 | 136,954 |
| Americas | 3,429 | 1,970 | 2,722 | 1,027 | 2,200 | 2,170 | 18,180 | 5,008 | 1,348 | 11,225 | 49,279 |
| Asia | 50,894 | 162,254 | 168,145 | 256,863 | 207,141 | 56,229 | 315,006 | 188,605 | 229,636 | 140,067 | 1,774,841 |
| Europe | 372 | 1,380 | 909 | 8,220 | 30 | 679 | 621 | 6,549 | 2,907 | 777 | 22,444 |
| Oceania | 1,849 | 5,158 | 5,914 | 2,682 | 652 | 1,230 | 328 | 151 | 7 | 32 | 18,003 |
| High human development | 2,220 | 6,044 | 8,247 | 11,192 | 2,238 | 1,090 | 1,858 | 4,706 | 984 | 798 | 39,377 |
| Medium human development | 42,523 | 147,331 | 165,961 | 233,338 | 196,087 | 55,386 | 317,237 | 194,881 | 229,087 | 150,410 | 1,732,241 |
| Low human development | 33,549 | 33,888 | 14,715 | 33,795 | 15,980 | 10,968 | 25,778 | 15,088 | 26,870 | 19,271 | 229,901 |
| **Total** | **78,292** | **187,262** | **188,923** | **278,325** | **214,305** | **67,444** | **344,873** | **214,675** | **256,941** | **170,478** | **2,001,519** |

Source: EM-DAT, CRED, University of Louvain, Belgium

Globally, an average of 200 million people a year were affected by disasters between 1992 and 2001. The greatest numbers of people affected during the decade were recorded in 1998: a devastating year during which floods in China, Hurricane Mitch in Central America and drought in Brazil, Cuba and Guyana affected millions. Throughout the decade, countries of medium human development (which include China and India, two disaster-prone countries with large populations) have consistently seen the highest numbers of people affected. They account for 86.6 per cent of the total for the decade.

chapter 8

**Table 4** Total amount of disaster estimated damage, by continent and by year (1992 to 2001) in millions of US dollars (2001 prices)

| | 1992 | 1993 | 1994 | 1995 | 1996 | 1997 | 1998 | 1999 | 2000 | 2001 | Total |
|---|---|---|---|---|---|---|---|---|---|---|---|
| Africa | 288 | 11 | 513 | 192 | 125 | 8 | 357 | 770 | 154 | 318 | 2,736 |
| Americas | 57,431 | 27,113 | 35,613 | 25,932 | 13,931 | 9,588 | 20,140 | 13,686 | 3,357 | 9,725 | 216,516 |
| Asia | 7,589 | 17,028 | 14,758 | 180,579 | 33,688 | 28,385 | 31,146 | 28,271 | 16,964 | 13,487 | 371,895 |
| Europe | 3,068 | 1,906 | 12,215 | 14,353 | 192 | 10,667 | 2,709 | 38,008 | 8,520 | 739 | 92,376 |
| Oceania | 2,571 | 1,746 | 2,028 | 1,483 | 1,057 | 257 | 194 | 904 | 519 | 141 | 10,901 |
| High human development | 62,279 | 27,308 | 49,176 | 192,319 | 12,907 | 19,695 | 15,129 | 41,472 | 11,108 | 7,546 | 438,940 |
| Medium human development | 7,278 | 19,799 | 15,219 | 11,686 | 32,424 | 28,953 | 36,947 | 40,145 | 11,522 | 16,359 | 220,331 |
| Low human development | 1,390 | 698 | 732 | 18,535 | 3,662 | 257 | 2,469 | 22 | 6,884 | 505 | 35,152 |
| **Total** | **70,947** | **47,804** | **65,127** | **222,539** | **48,993** | **48,905** | **54,546** | **81,639** | **29,514** | **24,410** | **694,424** |

*Source: EM-DAT, CRED, University of Louvain, Belgium*

Some 63 per cent of estimated damage is reported in nations of high human development, due to higher value infrastructure and greater insurance penetration which leads to better reporting. Nations of low human development reported just 5 per cent of estimated damage.
Damage estimations are notoriously unreliable. Methodologies are not standard and coverage is not complete. Depending on where the disaster occurred and who is reporting, estimates will vary from none to billions of US dollars. In order to take into account inflation, estimated damages have been recalculated at constant 2001 prices, using the US consumer prices index (US Department of Labor: http://stats.bls.gov/cpihome.htm).

**Table 5** Total number of reported disasters, by type of phenomenon and by year (1992 to 2001)

| | 1992 | 1993 | 1994 | 1995 | 1996 | 1997 | 1998 | 1999 | 2000 | 2001 | Total |
|---|---|---|---|---|---|---|---|---|---|---|---|
| Avalanches/landslides | 14 | 22 | 8 | 15 | 24 | 13 | 22 | 15 | 29 | 21 | 183 |
| Droughts/famines | 31 | 14 | 9 | 16 | 8 | 18 | 34 | 30 | 47 | 46 | 253 |
| Earthquakes | 24 | 14 | 22 | 25 | 11 | 14 | 16 | 33 | 30 | 26 | 215 |
| Extreme temperatures | 7 | 4 | 10 | 13 | 5 | 13 | 13 | 8 | 31 | 22 | 126 |
| Floods | 57 | 82 | 80 | 88 | 69 | 77 | 90 | 112 | 152 | 156 | 963 |
| Forest/scrub fires | 8 | 2 | 13 | 7 | 5 | 15 | 16 | 22 | 30 | 14 | 132 |
| Volcanic eruptions | 5 | 6 | 6 | 4 | 5 | 4 | 4 | 5 | 5 | 6 | 50 |
| Windstorms | 74 | 101 | 66 | 59 | 62 | 67 | 73 | 85 | 100 | 96 | 783 |
| Other natural disasters* | 1 | 4 | 1 | 4 | 2 | 3 | 2 | 2 | 4 | 2 | 25 |
| Subtotal hydro-meteorological disasters | 192 | 229 | 187 | 202 | 175 | 206 | 250 | 274 | 393 | 357 | 2,465 |
| Subtotal geophysical disasters | 29 | 20 | 28 | 29 | 16 | 18 | 20 | 38 | 35 | 32 | 265 |
| Total natural disasters | 221 | 249 | 215 | 231 | 191 | 224 | 270 | 312 | 428 | 389 | 2,730 |
| Industrial accidents | 24 | 36 | 34 | 39 | 35 | 35 | 43 | 36 | 50 | 54 | 386 |
| Miscellaneous accidents | 13 | 26 | 30 | 27 | 37 | 30 | 27 | 51 | 47 | 50 | 338 |
| Transport accidents | 110 | 113 | 127 | 104 | 135 | 136 | 145 | 212 | 234 | 219 | 1,535 |
| Total technological disasters | 147 | 175 | 191 | 170 | 207 | 201 | 215 | 299 | 331 | 323 | 2,259 |
| **Total** | **368** | **424** | **406** | **401** | **398** | **425** | **485** | **611** | **759** | **712** | **4,989** |

*Insect infestation and waves/surges

*Source: EM-DAT, CRED, University of Louvain, Belgium*

The number of technological disasters has increased over the decade from 147 in 1992 to 323 in 2001, with a peak of 329 in 2000. They now account for 45 per cent of all disasters recorded in EM-DAT. Floods and windstorms represent 64 per cent of the total of natural disasters (35 per cent of total disasters).

**Table 6** Total number of people reported killed by disasters, by type of phenomenon and by year (1992 to 2001)

| | 1992 | 1993 | 1994 | 1995 | 1996 | 1997 | 1998 | 1999 | 2000 | 2001 | Total |
|---|---|---|---|---|---|---|---|---|---|---|---|
| Avalanches/landslides | 1,070 | 1,548 | 280 | 1,497 | 1,129 | 801 | 994 | 351 | 1,099 | 692 | 9,461 |
| Droughts/famines | 2,571 | 0 | 0 | 54,000 | 54,000 | 54,530 | 57,875 | 54,029 | 370 | 199 | 277,574 |
| Earthquakes | 3,936 | 10,113 | 1,242 | 7,966 | 582 | 3,076 | 7,412 | 21,870 | 204 | 21,355 | 77,756 |
| Extreme temperatures | 388 | 106 | 416 | 1,730 | 300 | 619 | 3,225 | 771 | 922 | 1,653 | 10,130 |
| Floods | 5,367 | 5,930 | 6,504 | 7,525 | 8,040 | 6,602 | 11,186 | 34,366 | 6,307 | 4,680 | 96,507 |
| Forest/scrub fires | 122 | 3 | 84 | 29 | 45 | 32 | 109 | 70 | 47 | 33 | 574 |
| Volcanic eruptions | 2 | 99 | 101 | 0 | 4 | 53 | 0 | 0 | 0 | 0 | 259 |
| Windstorms | 1,355 | 2,965 | 4,064 | 3,774 | 3,649 | 5,330 | 24,552 | 11,899 | 1,129 | 1,730 | 60,447 |
| Other natural disasters* | 0 | 59 | 31 | 0 | 32 | 400 | 2,182 | 3 | 1 | 0 | 2,708 |
| Subtotal hydro-meteorological disasters | 10,873 | 10,611 | 11,379 | 68,555 | 67,195 | 68,314 | 100,123 | 101,489 | 9,875 | 8,987 | 457,401 |
| Subtotal geophysical disasters | 3,938 | 10,212 | 1,343 | 7,966 | 586 | 3,129 | 7,412 | 21,870 | 204 | 21,355 | 78,015 |
| Total natural disasters | 14,811 | 20,823 | 12,722 | 76,521 | 67,781 | 71,443 | 107,535 | 123,359 | 10,079 | 30,342 | 535,416 |
| Industrial accidents | 1,385 | 1,244 | 779 | 513 | 674 | 1,033 | 1,942 | 742 | 1,613 | 1,148 | 11,073 |
| Miscellaneous accidents | 321 | 1,103 | 1,719 | 1,660 | 1,159 | 1,277 | 652 | 1,330 | 1,112 | 1,792 | 12,125 |
| Transport accidents | 5,747 | 7,142 | 6,653 | 5,255 | 7,419 | 5,658 | 5,954 | 6,668 | 7,462 | 5,791 | 63,749 |
| Total technological disasters | 7,453 | 9,489 | 9,151 | 7,428 | 9,252 | 7,968 | 8,548 | 8,740 | 10,187 | 8,731 | 86,947 |
| **Total** | **22,264** | **30,312** | **21,873** | **83,949** | **77,033** | **79,411** | **116,083** | **132,099** | **20,266** | **39,073** | **622,363** |

*Insect infestation and waves/surges

Source: EM-DAT, CRED, University of Louvain, Belgium

Natural disasters accounted for 86 per cent of all fatalities during the period from 1992 to 2001, although 2000 saw technological disasters claim more lives for the first time.

**Table 7** Total number of people reported affected by disasters, by type of phenomenon and by year (1992 to 2001) in thousands

| | 1992 | 1993 | 1994 | 1995 | 1996 | 1997 | 1998 | 1999 | 2000 | 2001 | Total |
|---|---|---|---|---|---|---|---|---|---|---|---|
| Avalanches/landslides | 79 | 80 | 298 | 1,122 | 9 | 34 | 214 | 15 | 208 | 67 | 2,128 |
| Droughts/famines | 39,944 | 19,132 | 15,515 | 30,431 | 5,836 | 8,550 | 24,647 | 38,647 | 176,477 | 86,757 | 445,936 |
| Earthquakes | 787 | 270 | 730 | 3,029 | 1,996 | 593 | 1,878 | 3,893 | 2,458 | 19,307 | 34,942 |
| Extreme temperatures | 16 | 3,001 | 1,108 | 535 | 0 | 615 | 36 | 725 | 28 | 213 | 6,278 |
| Floods | 23,421 | 149,341 | 129,688 | 197,504 | 180,113 | 43,782 | 291,739 | 150,167 | 62,098 | 34,027 | 1,261,880 |
| Forest/scrub fires | 52 | 0 | 3,067 | 12 | 6 | 53 | 167 | 19 | 39 | 6 | 3,421 |
| Volcanic eruptions | 357 | 174 | 236 | 23 | 7 | 7 | 8 | 34 | 119 | 78 | 1,043 |
| Windstorms | 13,563 | 15,209 | 38,246 | 45,619 | 26,302 | 13,594 | 26,077 | 21,153 | 15,459 | 29,975 | 245,198 |
| Other natural disasters* | 0 | 0 | 0 | 0 | 0 | 29 | 10 | 1 | 17 | 0 | 58 |
| Subtotal hydro-meteorological disasters | 77,076 | 186,763 | 187,923 | 275,223 | 212,266 | 66,657 | 342,890 | 210,727 | 254,326 | 151,046 | 1,964,897 |
| Subtotal geophysical disasters | 1,144 | 443 | 966 | 3,052 | 2,003 | 601 | 1,885 | 3,928 | 2,577 | 19,386 | 35,985 |
| Total natural disasters | 78,219 | 187,207 | 188,889 | 278,276 | 214,269 | 67,258 | 344,776 | 214,655 | 256,903 | 170,431 | 2,000,882 |
| Industrial accidents | 18 | 8 | 19 | 27 | 16 | 163 | 63 | 3 | 17 | 19 | 353 |
| Miscellaneous accidents | 3 | 45 | 11 | 19 | 18 | 20 | 31 | 12 | 15 | 25 | 199 |
| Transport accidents | 52 | 2 | 3 | 3 | 3 | 3 | 4 | 5 | 6 | 3 | 85 |
| Total technological disasters | 73 | 56 | 34 | 50 | 36 | 186 | 97 | 21 | 38 | 47 | 637 |
| **Total** | **78,292** | **187,262** | **188,923** | **278,325** | **214,305** | **67,444** | **344,873** | **214,675** | **256,941** | **170,478** | **2,001,519** |

*Insect infestation and waves/surges

Source: EM-DAT, CRED, University of Louvain, Belgium

During the decade, hydro-meteorological disasters affected by far the greatest number of people, some 98 per cent of the total of people reported affected. Floods are the disasters that affect most people, accounting for 63 per cent of the total for 1992-2001.

**Table 8** Total amount of disaster estimated damage, by type of phenomenon and by year (1992 to 2001) in millions of US dollars (2001 prices)

| | 1992 | 1993 | 1994 | 1995 | 1996 | 1997 | 1998 | 1999 | 2000 | 2001 | Total |
|---|---|---|---|---|---|---|---|---|---|---|---|
| Avalanches/landslides | 544 | 874 | 75 | 12 | 0 | 18 | 0 | 0 | 173 | 61 | 1,757 |
| Droughts/famines | 3,141 | 1,359 | 1,469 | 6,701 | 1,356 | 450 | 489 | 7,262 | 6,494 | 3,783 | 32,505 |
| Earthquakes | 824 | 1,118 | 31,721 | 154,597 | 597 | 5,355 | 412 | 33,380 | 147 | 9,465 | 237,615 |
| Extreme temperatures | 3,717 | 0 | 780 | 968 | 0 | 3,304 | 4,033 | 1,060 | 128 | 0 | 13,990 |
| Floods | 7,061 | 30,061 | 23,101 | 30,244 | 27,977 | 12,781 | 32,866 | 14,112 | 10,991 | 2,755 | 191,950 |
| Forest/scrub fires | 531 | 1,230 | 182 | 156 | 1,945 | 18,710 | 615 | 520 | 1,084 | 70 | 25,043 |
| Volcanic eruptions | 0 | 1 | 480 | 1 | 19 | 9 | 0 | 0 | 0 | 4 | 513 |
| Windstorms | 54,119 | 10,863 | 5,959 | 28,268 | 14,124 | 8,241 | 15,936 | 25,088 | 9,929 | 8,263 | 180,790 |
| Other natural disasters * | 0 | 0 | 0 | 121 | 0 | 4 | 2 | 0 | 124 | 0 | 250 |
| Subtotal hydro-meteorological disasters | 69,113 | 44,388 | 31,566 | 66,470 | 45,402 | 43,509 | 53,942 | 48,042 | 28,923 | 14,932 | 446,286 |
| Subtotal geophysical disasters | 824 | 1,119 | 32,201 | 154,598 | 616 | 5,364 | 412 | 33,380 | 147 | 9,468 | 238,128 |
| Total natural disasters | 69,937 | 45,507 | 63,767 | 221,067 | 46,017 | 48,872 | 54,354 | 81,423 | 29,070 | 24,400 | 684,414 |
| Industrial accidents | 662 | 90 | 54 | 482 | 1,362 | 21 | 140 | 3 | 0 | 10 | 2,824 |
| Miscellaneous accidents | 35 | 1,424 | 707 | 210 | 1,381 | 0 | 20 | 2 | 444 | 0 | 4,224 |
| Transport accidents | 313 | 784 | 598 | 781 | 233 | 11 | 32 | 211 | 0 | 0 | 2,962 |
| Total technological disasters | 1,010 | 2,298 | 1,359 | 1,472 | 2,976 | 33 | 193 | 216 | 444 | 10 | 10,011 |
| **Total** | **70,947** | **47,804** | **65,127** | **222,539** | **48,993** | **48,905** | **54,546** | **81,639** | **29,514** | **24,410** | **694,424** |

Source: EM-DAT, CRED, University of Louvain, Belgium

*Insect infestation and waves/surges

Estimates of damage from natural disasters should be treated with caution, as damage to infrastructure will always result in a higher estimate than suffering of individuals, and the financial value attached to infrastructure in developed nations is much higher than that in developing countries. From 1992 to 2001, hydro-meteorological disasters accounted for 64 per cent of the decade's estimated damage from natural and technological disasters, compared to 34 per cent for geophysical disasters and 2 per cent for technological disasters.

**Table 9** Total number of reported disasters, by continent and by type of phenomenon (1992 to 2001)

| | Africa | Americas | Asia | Europe | Oceania | HHD¹ | MHD² | LHD³ | Total |
|---|---|---|---|---|---|---|---|---|---|
| Avalanches/landslides | 12 | 40 | 101 | 25 | 5 | 23 | 137 | 23 | 183 |
| Droughts/famines | 113 | 39 | 77 | 13 | 11 | 20 | 133 | 100 | 253 |
| Earthquakes | 10 | 48 | 112 | 37 | 8 | 42 | 152 | 21 | 215 |
| Extreme temperatures | 6 | 30 | 35 | 51 | 4 | 40 | 71 | 15 | 126 |
| Floods | 207 | 216 | 362 | 153 | 25 | 217 | 518 | 228 | 963 |
| Forest/scrub fires | 11 | 55 | 18 | 39 | 9 | 72 | 51 | 9 | 132 |
| Volcanic eruptions | 3 | 23 | 16 | 2 | 6 | 12 | 37 | 1 | 50 |
| Windstorms | 49 | 283 | 322 | 71 | 58 | 336 | 347 | 100 | 783 |
| Other natural disasters* | 4 | 4 | 14 | 1 | 2 | 1 | 17 | 7 | 25 |
| Subtotal hydro-meteorological disasters | 402 | 667 | 929 | 353 | 114 | 709 | 1,274 | 482 | 2,465 |
| Subtotal geophysical disasters | 13 | 71 | 128 | 39 | 14 | 54 | 189 | 22 | 265 |
| Total natural disasters | 415 | 738 | 1,057 | 392 | 128 | 763 | 1,463 | 504 | 2,730 |
| Industrial accidents | 37 | 55 | 225 | 67 | 2 | 66 | 276 | 44 | 386 |
| Miscellaneous accidents | 57 | 45 | 178 | 53 | 5 | 81 | 207 | 50 | 338 |
| Transport accidents | 437 | 233 | 668 | 186 | 11 | 200 | 920 | 415 | 1,535 |
| Total technological disasters | 531 | 333 | 1,071 | 306 | 18 | 347 | 1,403 | 509 | 2,259 |
| **Total** | **946** | **1,071** | **2,128** | **698** | **146** | **1,110** | **2,866** | **1,013** | **4,989** |

*Insect infestation and waves/surges

¹ High human development   ² Medium human development   ³ Low human development

*Source: EM-DAT, CRED, University of Louvain, Belgium*

Asia remains the continent most affected by both natural (39 per cent of the total number reported) and technological disasters (47 per cent of the total). In both Africa and Asia, technological disasters outnumber natural catastrophes.

chapter 8

## Table 10 Total number of people reported killed by disasters, by continent and by type of phenomenon (1992 to 2001)

| | Africa | Americas | Asia | Europe | Oceania | HHD¹ | MHD² | LHD³ | Total |
|---|---|---|---|---|---|---|---|---|---|
| Avalanches/landslides | 286 | 1,960 | 6,106 | 1,030 | 79 | 432 | 7,700 | 1,329 | 9,461 |
| Droughts/famines | 6,384 | 41 | 271,051 | 0 | 98 | 0 | 1,081 | 276,493 | 277,574 |
| Earthquakes | 784 | 3,463 | 52,440 | 20,998 | 71 | 8,316 | 62,020 | 7,420 | 77,756 |
| Extreme temperatures | 105 | 1,995 | 5,469 | 2,538 | 23 | 2,321 | 6,740 | 1,069 | 10,130 |
| Floods | 9,243 | 35,848 | 50,034 | 1,362 | 20 | 2,191 | 78,661 | 15,655 | 96,507 |
| Forest/scrub fires | 132 | 104 | 181 | 150 | 7 | 163 | 273 | 138 | 574 |
| Volcanic eruptions | 0 | 70 | 180 | 0 | 9 | 39 | 220 | 0 | 259 |
| Windstorms | 1,252 | 23,359 | 34,895 | 712 | 229 | 3,445 | 50,110 | 6,892 | 60,447 |
| Other natural disasters* | 0 | 15 | 511 | 0 | 2,182 | 0 | 2,707 | 1 | 2,708 |
| Subtotal hydro-meteorological disasters | 17,402 | 63,322 | 368,247 | 5,792 | 2,638 | 8,552 | 147,272 | 301,577 | 457,401 |
| Subtotal geophysical disasters | 784 | 3,533 | 52,620 | 20,998 | 80 | 8,355 | 62,240 | 7,420 | 78,015 |
| Total natural disasters | 18,186 | 66,855 | 420,867 | 26,790 | 2,718 | 16,907 | 209,512 | 308,997 | 535,416 |
| Industrial accidents | 2,551 | 638 | 6,654 | 1,208 | 22 | 326 | 8,073 | 2,674 | 11,073 |
| Miscellaneous accidents | 1,490 | 1,762 | 7,674 | 1,153 | 46 | 2,112 | 8,917 | 1,096 | 12,125 |
| Transport accidents | 17,849 | 10,038 | 28,486 | 6,843 | 533 | 8,119 | 34,631 | 20,999 | 63,749 |
| Total technological disasters | 21,890 | 12,438 | 42,814 | 9,204 | 601 | 10,557 | 51,621 | 24,769 | 86,947 |
| **Total** | **40,076** | **79,293** | **463,681** | **35,994** | **3,319** | **27,464** | **261,133** | **333,766** | **622,363** |

Source: EM-DAT, CRED, University of Louvain, Belgium

*Insect infestation and waves/surges

¹ High human development   ² Medium human development   ³ Low human development

More than half of the people killed by disaster in the years from 1992 to 2001 lived in countries of low human development. Worldwide, natural disasters claim 86 per cent of all disaster deaths, although in Africa technological disasters account for 55 per cent of disaster fatalities.

**Table 11** Total number of people reported affected by disasters, by continent and by type of phenomenon (1992 to 2001) in thousands

| | Africa | Americas | Asia | Europe | Oceania | HHD[1] | MHD[2] | LHD[3] | Total |
|---|---|---|---|---|---|---|---|---|---|
| Avalanches/landslides | 4 | 435 | 1,671 | 17 | 1 | 13 | 2,086 | 28 | 2,128 |
| Droughts/famines | 112,726 | 17,255 | 301,362 | 6,010 | 8,583 | 13,000 | 337,064 | 95,872 | 445,936 |
| Earthquakes | 145 | 3,518 | 28,799 | 2,436 | 45 | 2,446 | 31,119 | 1,378 | 34,942 |
| Extreme temperatures | 0 | 83 | 839 | 754 | 4,601 | 4,651 | 1,391 | 236 | 6,278 |
| Floods | 18,929 | 9,609 | 1,227,387 | 5,720 | 234 | 6,554 | 1,140,712 | 114,614 | 1,261,880 |
| Forest/scrub fires | 6 | 126 | 3,105 | 123 | 61 | 182 | 3,184 | 56 | 3,421 |
| Volcanic eruptions | 9 | 514 | 360 | 0 | 159 | 75 | 967 | 0 | 1,043 |
| Windstorms | 5,041 | 17,668 | 210,894 | 7,297 | 4,298 | 12,311 | 215,290 | 17,597 | 245,198 |
| Other natural disasters* | 0 | 1 | 47 | 0 | 10 | 0 | 17 | 41 | 58 |
| Subtotal hydro-meteorological disasters | 136,706 | 45,178 | 1,745,305 | 19,922 | 17,787 | 36,710 | 1,699,744 | 228,444 | 1,964,897 |
| Subtotal geophysical disasters | 154 | 4,032 | 29,159 | 2,436 | 204 | 2,521 | 32,086 | 1,378 | 35,985 |
| Total natural disasters | 136,860 | 49,210 | 1,774,463 | 22,358 | 17,990 | 39,231 | 1,731,830 | 229,822 | 2,000,882 |
| Industrial accidents | 3 | 56 | 218 | 76 | 0 | 122 | 227 | 4 | 353 |
| Miscellaneous accidents | 32 | 5 | 144 | 6 | 12 | 16 | 166 | 17 | 199 |
| Transport accidents | 58 | 7 | 15 | 4 | 0 | 9 | 19 | 58 | 85 |
| Total technological disasters | 93 | 69 | 377 | 86 | 12 | 146 | 412 | 79 | 637 |
| **Total** | **136,954** | **49,279** | **1,774,841** | **22,444** | **18,003** | **39,377** | **1,732,241** | **229,901** | **2,001,519** |

Source: EM-DAT, CRED, University of Louvain, Belgium

*Insect infestation and waves/surges
[1] High human development    [2] Medium human development    [3] Low human development

Floods in Asia alone represent 61 per cent of the total number of people reported affected by disaster. Some 86 per cent of people affected lived in nations of medium human development.

chapter 8

**Table 12** Total amount of disaster estimated damage, by continent and by type of phenomenon (1992-2001) in millions of US dollars (2001 prices)

| | Africa | Americas | Asia | Europe | Oceania | HHD[1] | MHD[2] | LHD[3] | Total |
|---|---|---|---|---|---|---|---|---|---|
| Avalanches/landslides | 0 | 1,290 | 439 | 28 | 0 | 284 | 1,473 | 0 | 1,757 |
| Droughts/famines | 417 | 3,516 | 12,688 | 10,787 | 5,096 | 16,387 | 15,755 | 363 | 32,505 |
| Earthquakes | 339 | 39,611 | 170,119 | 27,233 | 314 | 203,858 | 33,674 | 84 | 237,615 |
| Extreme temperatures | 1 | 8,419 | 4,497 | 1,073 | 0 | 9,494 | 4,496 | 0 | 13,990 |
| Floods | 967 | 36,422 | 119,167 | 34,532 | 862 | 66,067 | 99,452 | 26,431 | 191,950 |
| Forest/scrub fires | 4 | 3,641 | 20,941 | 200 | 257 | 3,930 | 21,102 | 12 | 25,043 |
| Volcanic eruptions | 0 | 10 | 1 | 22 | 480 | 31 | 482 | 0 | 513 |
| Windstorms | 828 | 117,899 | 41,508 | 16,800 | 3,755 | 132,132 | 40,767 | 7,891 | 180,790 |
| Other natural disasters* | 6 | 121 | 0 | 0 | 124 | 124 | 123 | 4 | 250 |
| Subtotal hydro-meteorological disasters | 2,222 | 171,309 | 199,241 | 63,420 | 10,093 | 228,418 | 183,168 | 34,700 | 446,286 |
| Subtotal geophysical disasters | 339 | 39,620 | 170,121 | 27,254 | 794 | 203,888 | 34,156 | 84 | 238,128 |
| Total natural disasters | 2,561 | 210,929 | 369,362 | 90,675 | 10,887 | 432,306 | 217,323 | 34,784 | 684,414 |
| Industrial accidents | 44 | 1,658 | 628 | 479 | 14 | 781 | 1,820 | 224 | 2,824 |
| Miscellaneous accidents | 5 | 2,899 | 731 | 589 | 0 | 4,139 | 81 | 5 | 4,224 |
| Transport accidents | 126 | 1,030 | 1,173 | 633 | 0 | 1,715 | 1,108 | 140 | 2,962 |
| Total technological disasters | 175 | 5,587 | 2,533 | 1,702 | 14 | 6,634 | 3,008 | 368 | 10,011 |
| **Total** | **2,736** | **216,516** | **371,895** | **92,376** | **10,901** | **438,940** | **220,331** | **35,152** | **694,424** |

Source: EM-DAT, CRED, University of Louvain, Belgium

*Insect infestation and waves/surges
[1] High human development   [2] Medium human development   [3] Low human development

Earthquakes and windstorms in high human development countries account for respectively 29 and 19 per cent of total amounts of estimated damage. In nations of low human development, estimated damage amounts to only 5 per cent of the total, while fatalities in these countries represent 54 per cent of the global number of deaths.

## Table 13 Total number of people reported killed and affected by disasters, by country (1982-1991; 1992-2001; and 2001)

| | Total number of people reported killed (1982-1991) | Total number of people reported affected (1982-1991) | Total number of people reported killed (1992-2001) | Total number of people reported affected (1992-2001) | Total number of people reported killed (2001) | Total number of people reported affected (2001) |
|---|---|---|---|---|---|---|
| **Africa** | **575,160** | **144,472,615** | **40,076** | **136,953,503** | **4,475** | **18,377,676** |
| Algeria | 359 | 48,956 | 1,432 | 145,327 | 921 | 50,423 |
| Angola | 280 | 3,681,000 | 1,132 | 181,246 | 321 | 40,047 |
| Benin | 64 | 3,149,000 | 111 | 835,283 | 26 | 7 |
| Botswana | 8 | 4,067,847 | 23 | 244,276 | – | – |
| Burkina Faso | 16 | 4,085,396 | 28 | 164,350 | – | – |
| Burundi | 112 | 3,600 | 18 | 1,143,620 | 12 | 224,710 |
| Cameroon | 1,847 | 798,537 | 721 | 5,445 | 34 | 1,000 |
| Canary Islands | 0 | 0 | 7 | 300 | 7 | 300 |
| Cape Verde | 32 | 7,722 | 18 | 16,306 | – | – |
| Central African Rep. | 31 | 0 | 22 | 79,628 | 3 | 3,000 |
| Chad | 3,093 | 3,682,212 | 131 | 1,232,436 | 129 | 319,244 |
| Comoros | 59 | 115,252 | 240 | 0 | – | – |
| Congo DR | 676 | 327,648 | 1,427 | 127,642 | 276 | 13,000 |
| Congo | 133 | 50 | 588 | 78,631 | 45 | 100 |
| Côte d'Ivoire | 99 | 7,070 | 406 | 280 | 19 | 2 |
| Djibouti | 10 | 260,300 | 145 | 585,775 | 0 | 95,000 |
| Egypt | 1,247 | 1,971 | 2,329 | 202,798 | 147 | 510 |
| Equatorial Guinea | 15 | 313 | 2 | 350 | 1 | 250 |
| Eritrea[6] | – | – | 133 | 2,688,725 | 0 | 738,000 |
| Ethiopia[6] | 301,139 | 40,865,940[6] | 812 | 33,341,851 | 42 | 1,039,558 |
| Gabon | 72 | 10,000 | 30 | 0 | – | – |
| Gambia, The | 0 | 0 | 154 | 38,750 | 1 | 250 |
| Ghana | 227 | 12,508,800 | 475 | 1,168,729 | 175 | 144,118 |
| Guinea | 292 | 21,436 | 677 | 226,242 | 115 | 220,135 |
| Guinea Bissau | 1 | 6,328 | 217 | 5,222 | – | – |
| Kenya | 626 | 617,154 | 1,759 | 15,013,200 | 147 | 4,400,070 |
| Lesotho | 40 | 680,000 | 1 | 503,751 | 1 | 2,001 |
| Liberia | 46 | 1,000,200 | 20 | 7,000 | 10 | – |
| Libya | 128 | 121 | 204 | 0 | – | – |
| Madagascar | 480 | 1,015,153 | 808 | 3,547,569 | 47 | 14 |
| Malawi | 506 | 5,743,701 | 123 | 16,647,279 | 74 | 508,754 |
| Mali | 121 | 1,826,635 | 3,736 | 12,263 | 2 | 3,500 |
| Mauritania | 27 | 3,310,200 | 2,392 | 767,106 | 11 | 20,361 |
| Mauritius | 161 | 44,858 | 5 | 3,300 | – | – |
| Morocco | 76 | 12,234 | 1,405 | 374,867 | 26 | 200 |

| | Total number of people reported killed (1982-1991) | Total number of people reported affected (1982-1991) | Total number of people reported killed (1992-2001) | Total number of people reported affected (1992-2001) | Total number of people reported killed (2001) | Total number of people reported affected (2001) |
|---|---|---|---|---|---|---|
| Mozambique | 105,810 | 8,578,508 | 1,463 | 8,875,293 | 89 | 849,326 |
| Namibia | 20 | 0 | 0 | 443,200 | – | – |
| Niger | 191 | 6,213,000 | 170 | 3,749,330 | 51 | 3,604,584 |
| Nigeria | 1,397 | 3,307,563 | 6,431 | 961,428 | 978 | 149,953 |
| Reunion | 65 | 10,261 | 16 | 600 | – | – |
| Rwanda | 285 | 501,678 | 216 | 1,271,763 | 17 | 25,000 |
| Sao Tome & Principe | 0 | 93,000 | 0 | 0 | – | – |
| Senegal | 0 | 1,218,000 | 387 | 441,870 | 18 | 353 |
| Seychelles | 0 | 0 | 5 | 1,237 | – | – |
| Sierra Leone | 172 | 0 | 965 | 200,025 | – | – |
| Somalia | 794 | 633,500 | 2,674 | 4,245,549 | 130 | 1,606,539 |
| South Africa | 1,707 | 5,272,372 | 2,165 | 417,940 | 206 | 36,245 |
| Sudan | 150,593 | 24,732,029 | 840 | 9,324,287 | 53 | 2,097,061 |
| Swaziland | 553 | 667,500 | 30 | 762,059 | 30 | 59 |
| Tanzania, UR | 385 | 2,366,577 | 1,758 | 8,949,129 | 86 | 1,096 |
| Togo | 0 | 400,000 | 3 | 281,905 | – | – |
| Tunisia | 205 | 128,500 | 34 | 89 | – | – |
| Uganda | 242 | 949,580 | 718 | 1,253,094 | 94 | 4,342 |
| Zambia | 429 | 800,000 | 246 | 5,734,019 | 118 | 1,448,564 |
| Zimbabwe | 289 | 700,913 | 224 | 10,651,139 | 13 | 730,000 |
| **Americas** | **60,147** | **62,298,697** | **79,293** | **49,278,660** | **3,449** | **11,224,660** |
| Anguilla | 0 | 0 | 0 | 150 | – | – |
| Antigua & Barbuda | 2 | 83,030 | 5 | 76,684 | – | – |
| Argentina | 280 | 12,429,698 | 488 | 802,151 | 64 | 256,950 |
| Bahamas | 100 | 0 | 5 | 3,200 | – | – |
| Barbados | 0 | 330 | 0 | 0 | – | – |
| Belize | 0 | 273 | 66 | 145,170 | 30 | 20,000 |
| Bermuda | 28 | 40 | 18 | 0 | – | – |
| Bolivia | 401 | 3,779,684 | 561 | 881,430 | 50 | 357,255 |
| Brazil | 3,795 | 29,315,099 | 1,964 | 12,067,601 | 123 | 1,015,038 |
| Canada | 508 | 80,266 | 483 | 574,835 | 11 | 1,230 |
| Chile | 895 | 2,258,838 | 293 | 370,613 | 28 | 14,980 |
| Colombia | 25,343 | 830,220 | 2,893 | 2,290,181 | 130 | 13,747 |
| Costa Rica | 109 | 359,654 | 125 | 815,289 | 0 | 1,437 |
| Cuba | 485 | 1,018,302 | 628 | 8,205,984 | 5 | 5,900,012 |
| Dominica | 2 | 10,710 | 15 | 3,891 | 3 | 175 |
| Dominican Rep. | 235 | 1,193,190 | 772 | 1,024,425 | – | – |
| Ecuador | 1,525 | 1,127,991 | 1,172 | 488,863 | 49 | 32,600 |

| | Total number of people reported killed (1982-1991) | Total number of people reported affected (1982-1991) | Total number of people reported killed (1992-2001) | Total number of people reported affected (1992-2001) | Total number of people reported killed (2001) | Total number of people reported affected (2001) |
|---|---|---|---|---|---|---|
| El Salvador | 1,755 | 910,254 | 1,863 | 2,106,211 | 1,159 | 2,004,190 |
| French Guiana | 0 | 0 | 0 | 70,000 | – | – |
| Grenada | 0 | 1,000 | 0 | 210 | – | – |
| Guadeloupe | 5 | 12,084 | 24 | 899 | 20 | – |
| Guatemala | 1,026 | 152,696 | 901 | 256,338 | 122 | 125,681 |
| Guyana | 0 | 281 | 10 | 797,200 | – | – |
| Haiti | 463 | 1,061,532 | 4,181 | 2,610,761 | 71 | 5,091 |
| Honduras | 409 | 114,137 | 15,258 | 3,666,540 | 21 | 852,445 |
| Jamaica | 119 | 1,427,640 | 14 | 5,372 | 1 | 200 |
| Martinique | 8 | 1,500 | 2 | 3,610 | – | – |
| Mexico | 11,995 | 833,322 | 4,716 | 2,771,055 | 43 | 6,400 |
| Montserrat | 11 | 12,040 | 32 | 13,000 | – | – |
| Netherlands Antilles | 0 | 0 | 17 | 40,004 | 15 | 4 |
| Nicaragua | 306 | 544,239 | 3,510 | 2,091,399 | 16 | 212,511 |
| Panama | 114 | 134,853 | 103 | 10,220 | 0 | 775 |
| Paraguay | 76 | 245,000 | 101 | 530,664 | – | – |
| Peru | 3,568 | 4,013,210 | 2,913 | 2,939,070 | 924 | 281,799 |
| Puerto Rico | 742 | 2,000 | 73 | 115,968 | 2 | 9,480 |
| St Kitts & Nevis | 1 | 1,330 | 5 | 12,980 | – | – |
| St Lucia | 45 | 3,000 | 4 | 1,125 | – | – |
| St Vincent & Grenadines | 0 | 1,360 | 3 | 300 | – | – |
| Suriname | 169 | 13 | 10 | 0 | 10 | – |
| Trinidad & Tobago | 6 | 1,020 | 5 | 610 | – | – |
| Turks & Caicos Is | 0 | 770 | – | – | – | – |
| United States | 4,961 | 229,183 | 5,402 | 2,781,391 | 515 | 107,657 |
| Uruguay | 0 | 20,000 | 116 | 28,087 | 0 | 5,000 |
| Venezuela | 660 | 78,908 | 30,531 | 665,176 | 37 | 3 |
| Virgin Islands (UK) | 0 | 10,000 | 0 | 3 | – | – |
| Virgin Islands (US) | 0 | 0 | 11 | 10,000 | – | – |
| **Asia** | **328,886** | **1,494,813,296** | **463,681** | **1,774,840,893** | **28,981** | **140,067,390** |
| Afghanistan | 9,526 | 485,969 | 9,475 | 7,040,314 | 353 | 4,000,270 |
| Armenia[1] | – | – | 106 | 1,604,810 | – | – |
| Azerbaijan[1] | – | – | 612 | 2,452,706 | – | – |
| Bahrain | 10 | 0 | 143 | 0 | – | – |
| Bangladesh | 166,882 | 246,186,789 | 8,208 | 71,772,943 | 469 | 729,033 |
| Bhutan | 0 | 0 | 222 | 1,600 | – | – |
| Cambodia | 100 | 900,000 | 1,094 | 13,336,614 | 56 | 1,989,182 |

| | Total number of people reported killed (1982-1991) | Total number of people reported affected (1982-1991) | Total number of people reported killed (1992-2001) | Total number of people reported affected (1992-2001) | Total number of people reported killed (2001) | Total number of people reported affected (2001) |
|---|---|---|---|---|---|---|
| China[7] | 22,624 | 498,473,051 | 32,672 | 977,674,128 | 2,063 | 61,491,112 |
| Georgia[1] | 293 | 266,745 | 349 | 1,220,480 | 0 | 523,900 |
| Hong Kong (China)[7] | 257 | 36,199 | 249 | 2,008,888 | – | – |
| India | 31,679 | 661,808,091 | 76,134 | 460,525,111 | 21,193 | 36,651,662 |
| Indonesia | 4,290 | 1,976,685 | 9,469 | 6,891,601 | 1,080 | 52,287 |
| Iran | 42,349 | 1,096,577 | 5,313 | 64,190,386 | 730 | 26,205,962 |
| Iraq | 796 | 500 | 113 | 808,007 | – | – |
| Israel | 67 | 398 | 125 | 2,159 | 15 | 247 |
| Japan | 2,100 | 3,279,278 | 6,609 | 2,772,870 | 118 | 21,380 |
| Jordan | 19 | 18,029 | 122 | 330,552 | 52 | – |
| Kazakhstan[1] | – | – | 247 | 644,216 | 3 | 3,680 |
| Korea, DPR | 545 | 20,071 | 270,675 | 9,998,967 | 124 | 187,584 |
| Korea, Rep. | 2,011 | 1,113,846 | 2,106 | 728,104 | 81 | 310,300 |
| Kuwait | 0 | 0 | 2 | 200 | – | – |
| Kyrgyzstan[1] | – | – | 286 | 264,328 | 20 | – |
| Lao, PDR | 14 | 1,102,315 | 223 | 2,940,552 | – | – |
| Lebanon | 65 | 1,500 | 35 | 104,102 | – | – |
| Macau | 0 | 0 | 0 | 3,986 | – | – |
| Malaysia | 505 | 103,943 | 719 | 74,660 | 19 | 18,788 |
| Maldives | 0 | 24,149 | 10 | 0 | – | – |
| Mongolia | 93 | 500,000 | 219 | 1,805,062 | 26 | 179,000 |
| Myanmar | 1,304 | 569,489 | 576 | 417,877 | 67 | 3,763 |
| Nepal | 2,075 | 918,152 | 3,633 | 931,794 | 154 | 21,026 |
| Oman | 0 | 0 | 0 | 0 | – | – |
| Pakistan | 3,106 | 1,443,112 | 7,730 | 27,312,592 | 386 | 1,315,211 |
| Palestine (West Bank) | 0 | 0 | 14 | 20 | – | – |
| Philippines | 24,819 | 36,276,615 | 7,016 | 58,092,847 | 682 | 2,398,869 |
| Saudi Arabia | 1,884 | 5,000 | 848 | 3,938 | 35 | – |
| Singapore | 24 | 0 | 3 | 1,437 | – | – |
| Sri Lanka | 718 | 10,119,636 | 580 | 4,424,013 | 13 | 1,000,200 |
| Syrian Arab Rep. | 0 | 0 | 209 | 658,288 | 54 | 191 |
| Taiwan (China) | 878 | 23,574 | 3,411 | 118,808 | 333 | 5,847 |
| Tajikistan[1] | – | – | 1,875 | 3,469,989 | 1 | 70,685 |
| Thailand | 2,676 | 7,558,930 | 2,359 | 21,056,825 | 282 | 470,563 |
| Turkmenistan[1] | – | – | 51 | 420 | – | – |
| United Arab Emirates | 112 | 0 | 111 | 116 | 14 | 15 |
| Uzbekistan[1] | – | – | 141 | 1,174,388 | 0 | 600,000 |
| Viet Nam | 5,002 | 18,866,113 | 8,535 | 27,697,582 | 462 | 1,816,616 |

| | Total number of people reported killed (1982-1991) | Total number of people reported affected (1982-1991) | Total number of people reported killed (1992-2001) | Total number of people reported affected (1992-2001) | Total number of people reported killed (2001) | Total number of people reported affected (2001) |
|---|---|---|---|---|---|---|
| Yemen[5] | 11 | 397,040 | 1,052 | 282,613 | 96 | 17 |
| Yemen, AR[5] | 1,545 | 551,500 | – | – | – | – |
| Yemen, PDR[5] | 507 | 690,000 | – | – | – | – |
| **Europe** | **40,577** | **6,641,289** | **35,994** | **22,443,714** | **2,159** | **776,927** |
| Albania | 186 | 3,212,806 | 15 | 46,500 | – | – |
| Austria | 95 | 30 | 253 | 10,194 | – | – |
| Azores | 172 | – | 74 | 1,215 | – | – |
| Belarus[1] | – | – | 61 | 63,468 | – | – |
| Belgium | 273 | 1,770 | 57 | 2,867 | – | – |
| Bosnia and Herzegovina[2] | – | – | 60 | 1,505 | – | – |
| Bulgaria | 94 | 3,160 | 44 | 6,759 | 2 | – |
| Croatia[2] | – | – | 137 | 3,425 | 0 | 1,200 |
| Cyprus | 0 | 0 | 59 | 4,307 | – | – |
| Czech Republic[3] | – | – | 47 | 102,111 | – | – |
| Czechoslovakia[3] | 94 | 95 | – | – | – | – |
| Denmark | 215 | 100 | 7 | 0 | – | – |
| Estonia[1] | – | – | 934 | 170 | 22 | 30 |
| Finland | 0 | 0 | 11 | 33 | – | – |
| France | 480 | 9,861 | 783 | 3,876,629 | 46 | 34,292 |
| Germany | 33 | 186 | 276 | 247,409 | 4 | 130 |
| Germany, Dem. Rep.[4] | 92 | – | – | – | – | – |
| Germany, Fed. Rep.[4] | 164 | 4,327 | – | – | – | – |
| Greece | 1,265 | 45,972 | 444 | 216,139 | 11 | 1,350 |
| Hungary | 4 | 0 | 188 | 145,774 | 81 | 10,000 |
| Iceland | 4 | 280 | 34 | 282 | – | – |
| Ireland | 364 | 0 | 32 | 4,200 | – | – |
| Italy | 914 | 58,436 | 836 | 237,526 | 139 | 7 |
| Latvia[1] | – | – | 21 | 0 | 21 | – |
| Lithuania[1] | – | – | 68 | 780,000 | 20 | – |
| Luxembourg | 0 | 0 | 0 | 0 | – | – |
| Macedonia, FYR[2] | – | – | 221 | 11,521 | 25 | 6 |
| Malta | 12 | 0 | 283 | 0 | – | – |
| Moldova, Rep.[1] | – | – | 59 | 2,654,537 | – | – |
| Netherlands | 6 | 0 | 176 | 265,871 | 12 | 180 |
| Norway | 236 | 0 | 270 | 6,130 | 0 | 1,500 |
| Poland | 228 | 18,813 | 811 | 240,767 | 297 | 15,000 |
| Portugal | 248 | 4,597 | 171 | 1,632 | 90 | 222 |

| | Total number of people reported killed (1982-1991) | Total number of people reported affected (1982-1991) | Total number of people reported killed (1992-2001) | Total number of people reported affected (1992-2001) | Total number of people reported killed (2001) | Total number of people reported affected (2001) |
|---|---|---|---|---|---|---|
| Romania | 326 | 21,705 | 413 | 235,987 | 41 | 14,905 |
| Russian Federation[1] | 127 | 46 | 6,629 | 2,238,721 | 995 | 394,079 |
| Serbia Montenegro[2] | – | – | 118 | 83,185 | – | – |
| Slovakia[3] | – | – | 73 | 48,015 | – | – |
| Slovenia[2] | – | – | 0 | 700 | – | – |
| Soviet Union[1] | 29,984 | 2,160,788 | – | – | – | – |
| Spain | 974 | 796,734 | 569 | 6,069,312 | 66 | 28 |
| Sweden | 23 | 100 | 76 | 184 | – | – |
| Switzerland | 68 | 2,400 | 107 | 6,520 | 35 | 9 |
| Turkey | 1,919 | 94,185 | 20,815 | 2,058,037 | 154 | 3,696 |
| Ukraine[1] | 32 | 0 | 492 | 2,481,944 | 85 | 300,073 |
| United Kingdom | 1,445 | 203,692 | 270 | 290,138 | 13 | 220 |
| Yugoslavia[2] | 500 | 1,206 | – | – | – | – |
| **Oceania** | **1,130** | **1,486,310** | **3,319** | **18,002,532** | **9** | **31,833** |
| American Samoa | 25 | 0 | 0 | 0 | – | – |
| Australia | 415 | 23,155 | 369 | 15,649,450 | 6 | 8,638 |
| Cook Islands | 6 | 2,000 | 19 | 1,644 | – | 744 |
| Fiji | 72 | 591,515 | 81 | 430,730 | 1 | – |
| French Polynesia | 17 | 5,050 | 13 | 511 | – | – |
| Guam | 0 | 502 | 229 | 12,033 | – | – |
| Kiribati | 0 | 0 | 0 | 84,000 | – | – |
| Marshall Islands | 0 | 6,000 | 0 | 0 | – | – |
| Micronesia, Fed. States | 5 | 203 | 0 | 28,800 | – | – |
| New Caledonia | 2 | – | 0 | – | – | – |
| New Zealand | 23 | 20,460 | 4 | 3,117 | – | – |
| Niue | 0 | 200 | 0 | 0 | – | – |
| Palau | 0 | 0 | 1 | 12,004 | – | – |
| Papua New Guinea | 343 | 57,000 | 2,524 | 1,632,707 | 0 | 201 |
| Samoa | 21 | 285,000 | 0 | 0 | – | – |
| Solomon Islands | 134 | 180,874 | 4 | 88,880 | – | – |
| Tokelau | 0 | 1,832 | 0 | 0 | – | – |
| Tonga | 8 | 149,617 | 0 | 23,021 | – | 16,450 |
| Tuvalu | 0 | 700 | 18 | 150 | – | – |
| Vanuatu | 58 | 157,702 | 52 | 35,465 | 2 | 5,800 |
| Wallis & Futuna Island | 1 | 4,500 | 5 | 20 | – | – |
| **Total** | **1,005,900** | **1,709,712,207** | **622,363** | **2,001,519,302** | **39,073** | **170,478,486** |

*Source: EM-DAT, CRED, University of Louvain, Belgium*

[1] Prior to 1991 Soviet Union is considered one country, after this date separate countries. The western former republics of the USSR (Belarus, Estonia, Latvia, Lithuania, Moldova, Russian Federation, Ukraine) are included in Europe; the southern former republics (Armenia, Azerbaijan, Georgia, Kazakhstan, Kyrgyzstan, Tajikistan, Turkmenistan, Uzbekistan) are included in Asia.

[2] Prior to 1992 Yugoslavia is considered one country, after this date separate countries: Bosnia and Herzegovina, Croatia, Serbia, Slovenia, Macedonia.

[3] Prior to 1993 Czechoslovakia is considered one country, after this date separate countries: Czech Republic and Slovakia.

[4] Prior to October 1990, Germany was divided into Federal and Democratic Republics, after this date considered as one country.

[5] Prior to May 1990, Yemen was divided into Arab and People's Democratic Republics, after this date considered one country.

[6] Prior to 1993, Ethiopia is considered one country, after this date separate countries: Eritrea and Ethiopia.

[7] Since July 1997, Hong Kong is included in China.

– No data has been entered in the EM-DAT for one of the following reasons:

   ▪ some countries did not report any data;

   ▪ some countries were not affected by any disasters;

   ▪ any disasters that affected these countries did not correspond to EM-DAT criteria (see introductory text to this chapter).

The number of people affected by disaster in 2001 (170 million) was lower than the decade's average of 200 million, as was the number of people killed: 39,073 compared to an average of 62,235 deaths per year in the decade from

## Table 14 Refugees and asylum seekers by country/territory of origin (1995 to 2001)

| | 1995 | 1996 | 1997 | 1998 | 1999 | 2000 | 2001 |
|---|---|---|---|---|---|---|---|
| **Africa** | **5,191,200** | **3,623,200** | **2,897,000** | **2,880,950** | **3,072,800** | **3,254,300** | **2,923,000** |
| Algeria | – | – | – | 3,000 | 5,000 | – | 10,000 |
| Angola | 313,000 | 220,000 | 223,000 | 303,300 | 339,300 | 400,000 | 400,000 |
| Burundi | 290,000 | 285,000 | 248,000 | 281,000 | 311,000 | 421,000 | 370,000 |
| Cameroon | – | – | – | – | – | – | 2,000 |
| Central African Republic | – | – | – | – | – | – | 20,000 |
| Chad | 16,000 | 15,500 | 12,000 | 15,000 | 13,000 | 53,000 | 35,000 |
| Congo, DR of* | 58,600 | 116,800 | 132,000 | 136,000 | 229,000 | 342,000 | 350,000 |
| Congo, PR of | – | – | 40,000 | 20,000 | 25,000 | 22,000 | 30,000 |
| Djibouti | 10,000 | 10,000 | 5,000 | 3,000 | 1,000 | 1,000 | – |
| Egypt | – | – | – | – | 3,000 | – | – |
| Eritrea | 342,500 | 343,100 | 323,000 | 323,100 | 323,100 | 356,400 | 300,000 |
| Ethiopia | 110,700 | 58,000 | 48,000 | 39,600 | 53,300 | 36,200 | 15,000 |
| Ghana | – | 10,000 | 12,000 | 11,000 | 10,000 | 10,000 | 10,000 |
| Guinea | – | – | – | – | – | – | 5,000 |
| Guinea-Bissau | – | – | – | 11,150 | 5,300 | 1,500 | – |
| Kenya | – | – | – | 8,000 | 5,000 | – | – |
| Liberia | 725,000 | 755,000 | 475,000 | 310,000 | 249,000 | 196,000 | 215,000 |
| Mali | 90,000 | 80,000 | 16,000 | 3,000 | 2,000 | – | 3,000 |
| Mauritania | 80,000 | 65,000 | 55,000 | 30,000 | 45,000 | 45,000 | 50,000 |
| Mozambique | 97,000 | – | – | – | – | – | – |
| Namibia | – | – | – | – | 1,000 | – | 2,000 |
| Niger | 20,000 | 15,000 | 10,000 | – | – | – | – |
| Nigeria | – | – | – | – | – | – | 10,000 |
| Rwanda | 1,545,000 | 257,000 | 43,000 | 12,000 | 27,000 | 52,000 | 40,000 |
| Senegal | 17,000 | 17,000 | 17,000 | 10,000 | 10,000 | 10,000 | 10,000 |
| Sierra Leone | 363,000 | 350,000 | 297,000 | 480,000 | 454,000 | 419,000 | 180,000 |
| Somalia | 480,300 | 467,100 | 486,000 | 414,600 | 415,600 | 370,000 | 270,000 |
| Sudan | 448,100 | 433,700 | 353,000 | 352,200 | 423,200 | 392,200 | 470,000 |
| Togo | 95,000 | 30,000 | 6,000 | 3,000 | 3,000 | 2,000 | 1,000 |
| Uganda | 10,000 | 15,000 | 10,000 | 12,000 | 15,000 | 20,000 | 15,000 |
| Western Sahara | 80,000 | 80,000 | 86,000 | 100,000 | 105,000 | 105,000 | 110,000 |
| **East Asia and Pacific** | **640,950** | **648,200** | **723,000** | **763,200** | **864,100** | **1,056,000** | **1,078,500** |
| Cambodia | 26,300 | 34,400 | 77,000 | 51,000 | 15,100 | 16,400 | 16,000 |
| China (Tibet) | 141,000 | 128,000 | 128,000 | 128,000 | 130,000 | 130,000 | 130,000 |
| East Timor | – | – | – | – | 120,000 | 120,000 | 80,000 |
| Indonesia | 9,500 | 10,000 | 8,000 | 8,000 | 8,000 | 6,150 | 5,500 |
| Korea, DPR of | – | – | – | – | – | 50,000 | 50,000 |
| Laos | 8,900 | 3,500 | 14,000 | 12,100 | 13,900 | 400 | – |
| Myanmar | 160,400 | 184,300 | 215,000 | 238,100 | 240,100 | 380,250 | 438,000 |
| Philippines | – | – | – | 45,000 | 45,000 | 57,000 | 57,000 |
| Viet Nam | 294,850 | 288,000 | 281,000 | 281,000 | 292,000 | 295,800 | 302,000 |

| | 1995 | 1996 | 1997 | 1998 | 1999 | 2000 | 2001 |
|---|---|---|---|---|---|---|---|
| **South and central Asia** | **2,809,400** | **3,184,100** | **2,966,000** | **2,928,700** | **2,906,750** | **3,832,700** | **4,961,500** |
| Afghanistan | 2,328,400 | 2,628,550 | 2,622,000 | 2,628,600 | 2,561,050 | 3,520,350 | 4,600,000 |
| Bangladesh | 48,000 | 53,000 | 40,000 | – | – | – | 7,500 |
| Bhutan | 118,600 | 121,800 | 113,000 | 115,000 | 125,000 | 124,000 | 126,000 |
| India | – | 13,000 | 13,000 | 15,000 | 15,000 | 17,000 | 17,000 |
| Kazakhstan | – | – | – | – | – | 100 | – |
| Pakistan | – | – | – | – | – | – | 11,000 |
| Sri Lanka | 96,000 | 100,150 | 100,000 | 110,000 | 110,000 | 110,000 | 144,000 |
| Tajikistan | 170,400 | 215,600 | 32,000 | 15,100 | 62,500 | 59,750 | 56,000 |
| Uzbekistan | 48,000 | 52,000 | 46,000 | 45,000 | 33,200 | 1,500 | – |
| **Middle East** | **3,958,500** | **4,373,100** | **4,304,000** | **4,397,700** | **3,987,050** | **5,426,500** | **4,428,000** |
| Iran | 49,500 | 46,100 | 35,000 | 30,800 | 31,200 | 30,600 | 40,000 |
| Iraq | 622,900 | 608,500 | 526,000 | 555,800 | 534,450 | 409,300 | 265,000 |
| Lebanon | – | – | – | – | – | 4,400 | – |
| Palestinians | 3,286,100 | 3,718,500 | 3,743,000 | 3,811,100 | 3,931,400 | 4,982,100 | 4,123,000 |
| Syria | – | – | – | – | – | 100 | – |
| **Europe** | **1,805,600** | **1,875,150** | **1,343,100** | **1,241,300** | **1,238,100** | **755,900** | **674,000** |
| Albania | – | – | – | – | – | – | 4,200 |
| Armenia | 185,000 | 197,000 | 188,000 | 180,000 | 188,400 | – | 8,600 |
| Azerbaijan | 390,000 | 238,000 | 218,000 | 218,000 | 230,000 | – | 3,500 |
| Belarus | – | – | – | – | – | – | 2,800 |
| Bosnia and Herzegovina | 1,006,450 | 577,000 | 342,600 | 80,350 | 250,000 | 234,600 | 188,000 |
| Croatia | 200,000 | 300,000 | 335,000 | 329,000 | 336,000 | 314,700 | 272,000 |
| Georgia | 105,000 | 105,000 | 11,000 | 23,000 | 2,800 | 22,400 | 21,100 |
| Macedonia | – | – | – | – | – | – | 43,600 |
| Russian Federation | – | – | – | 500 | 12,350 | 22,700 | 30,200 |
| Slovenia | – | – | – | – | – | 4,400 | – |
| Turkey | 15,000 | 15,000 | 11,000 | 11,300 | 11,800 | 12,600 | 40,000 |
| Yugoslavia | 5,100 | 13,700 | 3,100 | 136,900 | 376,400 | 148,900 | 60,000 |
| **Americas and Caribbean** | **65,700** | **61,900** | **521,300** | **442,550** | **393,800** | **366,750** | **428,000** |
| Colombia | 200 | 0 | 300 | 600 | – | 2,300 | 30,000 |
| Cuba | 4,000 | 850 | 600 | 300 | 850 | 1,200 | 3,000 |
| El Salvador | 12,400 | 12,000 | 12,000 | 250,150 | 253,000 | 235,500 | 217,000 |
| Guatemala | 34,150 | 34,650 | 30,000 | 251,300 | 146,000 | 102,600 | 129,000 |
| Haiti | 1,500 | – | – | 600 | 23,000 | 20,600 | 25,000 |
| Mexico | – | – | – | – | – | – | 11,000 |
| Nicaragua | 16,150 | 18,200 | 19,000 | 18,000 | 18,000 | 3,800 | 13,000 |
| Peru | – | – | – | 350 | 1,700 | 750 | – |
| **Total** | **14,479,850** | **13,769,450** | **12,295,000** | **12,733,150** | **12,511,350** | **14,692,150** | **14,493,000** |

Notes: - indicates zero or near zero; * formerly Zaire.　　　　　　　　*Source: US Committee for Refugees*

More than half of the world's refugees and asylum seekers in 2001 were either Palestinians or Afghans. The number of refugees in need of protection from Afghanistan, Central African Republic, Colombia, Congo, Macedonia, Myanmar, Sri Lanka and other countries increased during the year, while the number of refugees from Bosnia and Herzegovina, Burundi, East Timor, Kosovo (Yugoslavia), Sierra Leone, Somalia and elsewhere declined. However, in part because asylum states do not always report country-of-origin data, this table understates the number of refugees and asylum seekers from many countries.

## Table 15 Refugees and asylum seekers by host country/territory (1995 to 2001)

| | 1995 | 1996 | 1997 | 1998 | 1999 | 2000 | 2001 |
|---|---|---|---|---|---|---|---|
| **Africa** | **5,222,300** | **3,682,700** | **2,944,000** | **2,924,000** | **3,147,000** | **3,346,000** | **3,052,000** |
| Algeria | 120,000 | 114,000 | 104,000 | 84,000 | 84,000 | 85,000 | 85,000 |
| Angola | 10,900 | 9,300 | 9,000 | 10,000 | 15,000 | 12,000 | 12,000 |
| Benin | 25,000 | 11,000 | 3,000 | 3,000 | 3,000 | 4,000 | 5,000 |
| Botswana | – | – | – | – | 1,000 | 3,000 | 4,000 |
| Burkina Faso | 21,000 | 26,000 | 2,000 | – | – | – | 1,000 |
| Burundi | 140,000 | 12,000 | 12,000 | 5,000 | 2,000 | 6,000 | 30,000 |
| Cameroon | 2,000 | 1,000 | 1,000 | 3,000 | 10,000 | 45,000 | 30,000 |
| Central African Rep. | 34,000 | 36,400 | 38,000 | 47,000 | 55,000 | 54,000 | 50,000 |
| Chad | – | – | – | 10,000 | 20,000 | 20,000 | 15,000 |
| Congo, DR of¹ | 1,332,000 | 455,000 | 255,000 | 220,000 | 235,000 | 276,000 | 270,000 |
| Congo, PR of | 15,000 | 16,000 | 21,000 | 20,000 | 40,000 | 126,000 | 100,000 |
| Côte d'Ivoire | 290,000 | 320,000 | 202,000 | 128,000 | 135,000 | 94,000 | 100,000 |
| Djibouti | 25,000 | 22,000 | 22,000 | 23,000 | 23,000 | 22,000 | 22,000 |
| Egypt | 10,400 | 46,000 | 46,000 | 46,000 | 47,000 | 57,000 | 73,000 |
| Eritrea | – | – | 3,000 | 3,000 | 2,000 | 1,000 | 2,000 |
| Ethiopia | 308,000 | 328,000 | 313,000 | 251,000 | 246,000 | 194,000 | 115,000 |
| Gabon | 1,000 | 1,000 | 1,000 | 1,000 | 15,000 | 15,000 | 20,000 |
| Gambia | 5,000 | 5,000 | 8,000 | 13,000 | 25,000 | 15,000 | 15,000 |
| Ghana | 85,000 | 35,000 | 20,000 | 15,000 | 12,000 | 13,000 | 10,000 |
| Guinea | 640,000 | 650,000 | 430,000 | 514,000 | 453,000 | 390,000 | 190,000 |
| Guinea-Bissau | 15,000 | 15,000 | 4,000 | 5,000 | 5,000 | 6,000 | 7,000 |
| Kenya | 225,000 | 186,000 | 196,000 | 192,000 | 254,000 | 233,000 | 215,000 |
| Liberia | 120,000 | 100,000 | 100,000 | 120,000 | 90,000 | 70,000 | 60,000 |
| Libya | 28,100 | 27,200 | 27,000 | 28,000 | 11,000 | 11,000 | 30,000 |
| Malawi | 2,000 | – | – | – | – | – | 5,000 |
| Mali | 15,000 | 15,000 | 17,000 | 5,000 | 7,000 | 7,000 | 10,000 |
| Mauritania | 35,000 | 15,000 | 5,000 | 20,000 | 25,000 | 25,000 | 25,000 |
| Mozambique | – | – | – | – | 1,000 | 2,000 | 5,000 |
| Namibia | 1,000 | 1,000 | 1,000 | 2,000 | 8,000 | 20,000 | 30,000 |
| Niger | 17,000 | 27,000 | 7,000 | 3,000 | 2,000 | 1,000 | 1,000 |
| Nigeria | 8,000 | 8,000 | 9,000 | 5,000 | 7,000 | 10,000 | 7,000 |
| Rwanda | – | 20,000 | 28,000 | 36,000 | 36,000 | 29,000 | 35,000 |
| Senegal | 68,000 | 51,000 | 41,000 | 30,000 | 42,000 | 41,000 | 45,000 |
| Sierra Leone | 15,000 | 15,000 | 15,000 | 10,000 | 7,000 | 3,000 | 15,000 |
| South Africa | 90,000 | 22,500 | 28,000 | 29,000 | 40,000 | 30,000 | 22,000 |
| Sudan | 450,000 | 395,000 | 365,000 | 360,000 | 363,000 | 385,000 | 370,000 |
| Swaziland | – | – | – | – | – | – | 1,000 |
| Tanzania | 703,000 | 335,000 | 295,000 | 329,000 | 413,000 | 543,000 | 500,000 |
| Togo | 10,000 | 10,000 | 12,000 | 11,000 | 10,000 | 11,000 | 10,000 |
| Tunisia | 500 | 300 | – | – | – | – | – |

| | 1995 | 1996 | 1997 | 1998 | 1999 | 2000 | 2001 |
|---|---|---|---|---|---|---|---|
| Uganda | 230,000 | 225,000 | 185,000 | 185,000 | 197,000 | 230,000 | 230,000 |
| Zambia | 125,400 | 126,000 | 118,000 | 157,000 | 205,000 | 255,000 | 270,000 |
| Zimbabwe | – | 1,000 | 1,000 | 1,000 | 1,000 | 2,000 | 10,000 |
| **East Asia and Pacific** | **452,850** | **449,600** | **535,100** | **559,200** | **657,300** | **791,700** | **815,800** |
| Australia | 7,500 | 7,400 | 18,000 | 15,000 | 17,000 | 16,700 | 21,800 |
| Cambodia | – | – | – | 200 | 100 | 50 | 1,000 |
| China[2] | 294,100 | 294,100 | 281,800 | 281,800 | 292,800 | 350,000 | 345,000 |
| Hong Kong[2] | 1,900 | 1,300 | n.a. | n.a. | n.a. | n.a. | n.a. |
| Indonesia | – | – | 100 | 100 | 120,000 | 120,800 | 81,300 |
| Japan | 9,900 | 300 | 300 | 500 | 400 | 3,800 | 6,400 |
| Korea, Rep. of | – | – | – | – | – | 350 | – |
| Malaysia[3] | 5,300 | 5,200 | 5,200 | 50,600 | 45,400 | 57,400 | 57,000 |
| Nauru | – | – | – | – | – | – | 1,100 |
| New Zealand | – | – | – | – | – | 3,100 | 3,100 |
| Papua New Guinea | 9,500 | 10,000 | 8,200 | 8,000 | 8,000 | 6,000 | 5,400 |
| Philippines | 450 | 50 | 100 | 300 | 200 | 200 | 200 |
| Solomon Islands | 1,000 | 1,000 | 800 | – | – | – | – |
| Thailand | 98,200 | 95,850 | 205,600 | 187,700 | 158,400 | 217,300 | 277,000 |
| Viet Nam | 25,000 | 34,400 | 15,000 | 15,000 | 15,000 | 16,000 | 16,000 |
| **South and central Asia** | **1,386,300** | **1,794,800** | **1,743,000** | **1,708,700** | **1,689,000** | **2,655,600** | **2,880,100** |
| Afghanistan | 18,400 | 18,900 | – | – | – | – | – |
| Bangladesh | 55,000 | 40,000 | 40,100 | 53,100 | 53,100 | 121,600 | 122,000 |
| India | 319,200 | 352,200 | 323,500 | 292,100 | 292,000 | 290,000 | 330,000 |
| Kazakhstan | 6,500 | 14,000 | 14,000 | 4,100 | 14,800 | 20,000 | 20,000 |
| Kyrgyzstan | 7,600 | 17,000 | 15,500 | 15,000 | 10,900 | 11,000 | 9,700 |
| Nepal | 106,600 | 109,800 | 116,000 | 118,000 | 130,000 | 129,000 | 131,000 |
| Pakistan | 867,500 | 1,215,700 | 215,650 | 1,217,400 | 1,127,000 | 2,019,000 | 2,200,000 |
| Tajikistan | 2,500 | 2,200 | 3,800 | 5,500 | 4,700 | 12,400 | 15,400 |
| Turkmenistan | – | 22,000 | 13,000 | 500 | 18,500 | 14,200 | 14,000 |
| Uzbekistan | 3,000 | 3,000 | 1,250 | 3,000 | 38,000 | 38,400 | 38,000 |
| **Middle East** | **5,499,100** | **5,840,550** | **5,708,000** | **5,814,100** | **5,849,000** | **6,035,300** | **6,832,900** |
| Gaza Strip | 683,600 | 716,900 | 746,000 | 773,000 | 798,400 | 824,600 | 852,600 |
| Iran | 2,075,500 | 2,020,000 | 1,900,000 | 1,931,000 | 1,835,000 | 1,895,000 | 2,558,300 |
| Iraq | 115,200 | 114,400 | 110,000 | 104,000 | 129,400 | 127,700 | 128,100 |
| Israel | – | – | – | – | 400 | 4,700 | 4,700 |
| Jordan | 1,294,800 | 1,362,500 | 1,413,800 | 1,463,800 | 1,518,000 | 1,580,000 | 1,643,900 |
| Kuwait | 55,000 | 42,000 | 90,000 | 52,000 | 52,000 | 52,000 | 50,000 |
| Lebanon | 348,300 | 355,100 | 362,300 | 368,300 | 378,100 | 383,200 | 389,600 |
| Saudi Arabia | 13,200 | 257,850 | 116,750 | 128,300 | 128,600 | 128,500 | 128,500 |
| Syria | 342,300 | 384,400 | 361,000 | 369,800 | 379,200 | 389,000 | 397,600 |
| United Arab Emirates | 400 | 400 | 500 | 200 | – | – | – |

| | 1995 | 1996 | 1997 | 1998 | 1999 | 2000 | 2001 |
|---|---|---|---|---|---|---|---|
| West Bank | 517,400 | 532,400 | 543,000 | 555,000 | 569,700 | 583,000 | 607,800 |
| Yemen | 53,400 | 54,600 | 64,900 | 68,700 | 60,000 | 67,600 | 71,800 |
| **Europe** | **2,520,700** | **2,479,100** | **2,020,300** | **1,728,400** | **1,909,100** | **1,153,300** | **960,500** |
| Albania | – | – | – | 25,000 | 5,000 | 500 | 400 |
| Armenia | 304,000 | 150,000 | 219,150 | 229,000 | 240,000 | – | 11,000 |
| Austria | 55,900 | 80,000 | 11,400 | 16,500 | 16,600 | 6,100 | 10,600 |
| Azerbaijan | 238,000 | 249,150 | 244,100 | 235,300 | 222,000 | 3,600 | 5,500 |
| Belarus | 7,000 | 10,800 | 33,500 | 16,500 | 2,900 | 3,200 | 3,100 |
| Belgium | 16,400 | 18,200 | 14,100 | 25,800 | 42,000 | 46,400 | 40,000 |
| Bosnia and Herzegovina | – | – | 40,000 | 40,000 | 60,000 | 38,200 | 36,000 |
| Bulgaria | 500 | 550 | 2,400 | 2,800 | 2,800 | 3,000 | 1,500 |
| Croatia | 189,500 | 167,000 | 50,000 | 27,300 | 24,000 | 22,500 | 21,900 |
| Cyprus | – | – | – | 200 | 300 | 300 | 1,600 |
| Czech Republic | 2,400 | 2,900 | 700 | 2,400 | 1,800 | 4,800 | 10,500 |
| Denmark | 9,600 | 24,600 | 13,000 | 6,100 | 8,500 | 10,300 | 10,000 |
| Finland | 750 | 1,700 | 1,600 | 2,300 | 3,800 | 2,600 | 2,000 |
| France | 30,000 | 29,200 | 16,000 | 17,400 | 30,000 | 26,200 | 13,000 |
| Georgia | – | – | 100 | 300 | 5,200 | 7,600 | 7,900 |
| Germany | 442,700 | 436,400 | 277,000 | 198,000 | 285,000 | 180,000 | 116,200 |
| Greece | 1,300 | 5,600 | 2,100 | 2,800 | 7,500 | 800 | 2,500 |
| Hungary | 9,100 | 5,400 | 3,400 | 3,200 | 6,000 | 4,200 | 2,900 |
| Iceland | – | – | – | – | 100 | 50 | – |
| Ireland | – | 1,800 | 4,300 | 6,900 | 8,500 | 7,700 | 9,500 |
| Italy | 60,700 | 10,600 | 20,000 | 6,800 | 24,900 | 13,700 | 14,000 |
| Latvia | 150 | – | – | – | – | – | – |
| Lithuania | 400 | – | 100 | 100 | 100 | 150 | 300 |
| Macedonia, FYR of | 7,000 | 5,100 | 3,500 | 7,300 | 17,400 | 9,000 | 3,500 |
| Netherlands | 39,300 | 46,200 | 64,200 | 47,000 | 40,000 | 29,600 | 31,000 |
| Norway | 11,200 | 12,700 | 3,100 | 2,500 | 9,500 | 8,600 | 10,000 |
| Poland | 800 | 3,200 | 1,200 | 1,300 | 1,300 | 2,300 | 100 |
| Portugal | 350 | 200 | 150 | 1,400 | 1,700 | 1,600 | 1,300 |
| Romania | 1,300 | 600 | 2,000 | 900 | 900 | 2,100 | 200 |
| Russian Federation | 500,000 | 484,000 | 324,000 | 161,900 | 104,300 | 36,200 | 28,300 |
| Slovak Republic | 1,600 | 2,000 | 100 | 300 | 400 | 400 | 3,100 |
| Slovenia | 24,000 | 10,300 | 5,300 | 7,300 | 5,000 | 12,000 | 2,700 |
| Spain | 4,300 | 7,200 | 3,300 | 2,500 | 4,500 | 1,100 | 1,000 |
| Sweden | 12,300 | 60,500 | 8,400 | 16,700 | 20,200 | 18,500 | 18,000 |
| Switzerland | 29,000 | 41,700 | 34,100 | 40,000 | 104,000 | 62,600 | 57,600 |
| Turkey | 21,150 | 13,000 | 5,000 | 12,000 | 9,100 | 9,900 | 10,100 |
| Ukraine | 6,000 | 8,000 | 4,900 | 8,600 | 5,800 | 5,500 | 5,500 |
| United Kingdom | 44,000 | 40,500 | 58,100 | 74,000 | 112,000 | 87,800 | 67,700 |
| Yugoslavia | 450,000 | 550,000 | 550,000 | 480,000 | 476,000 | 484,200 | 400,000 |

| | 1995 | 1996 | 1997 | 1998 | 1999 | 2000 | 2001 |
|---|---|---|---|---|---|---|---|
| **Americas and the Caribbean** | **256,400** | **232,800** | **616,000** | **739,950** | **737,000** | **562,100** | **596,700** |
| Argentina | – | 400 | 10,700 | 1,100 | 3,300 | 1,000 | 3,100 |
| Bahamas | 200 | – | 50 | 100 | 100 | 100 | 100 |
| Belize | 8,650 | 8,700 | 4,000 | 3,500 | 3,000 | 1,700 | – |
| Bolivia | 600 | 550 | 300 | 350 | 400 | – | 400 |
| Brazil | 2,000 | 2,200 | 2,300 | 2,400 | 2,300 | 2,700 | 4,000 |
| Canada | 24,900 | 26,100 | 48,800 | 46,000 | 53,000 | 54,400 | 70,000 |
| Chile | 300 | 200 | 300 | 100 | 300 | 300 | 500 |
| Colombia | 400 | 200 | 200 | 200 | 250 | 250 | 200 |
| Costa Rica | 20,500 | 23,150 | 23,100 | 23,100 | 22,900 | 7,300 | 10,700 |
| Cuba | 1,800 | 1,650 | 1,500 | 1,100 | 1,000 | 1,000 | 1,000 |
| Dominican Republic | 900 | 600 | 600 | 600 | 650 | 500 | 500 |
| Ecuador | 100 | 200 | 200 | 250 | 350 | 1,600 | 4,300 |
| El Salvador | 150 | 150 | 100 | 100 | – | – | – |
| Guatemala | 2,500 | 1,200 | 1,300 | 800 | 750 | 700 | 700 |
| Honduras | 50 | – | – | 100 | – | – | – |
| Jamaica | – | – | – | 50 | 50 | 50 | – |
| Mexico | 38,500 | 34,450 | 30,000 | 7,500 | 8,500 | 6,500 | 6,200 |
| Nicaragua | 450 | 900 | 700 | 150 | 500 | 300 | – |
| Panama | 800 | 650 | 300 | 1,300 | 600 | 1,300 | 1,300 |
| Peru | 700 | 300 | – | – | 700 | 750 | 700 |
| United States | 152,200 | 129,600 | 491,000 | 651,000 | 638,000 | 481,500 | 492,500 |
| Uruguay | – | – | – | – | 150 | 50 | 100 |
| Venezuela | 700 | 1,600 | 300 | 150 | 200 | 100 | 400 |
| **Total** | **16,266,800** | **15,337,650** | **14,479,550** | **13,566,400** | **13,988,000** | **14,543,700** | **15,138,000** |

Notes: – indicates zero or near zero; n.a. not available, or reported estimates unreliable; [1] formerly Zaire; [2] as of 1997, figures for Hong Kong are included in total for China; [3] USCR reclassified as refugees 45,00 Filipino Muslims from the island of Mindanao previously regarded as "refugee-like". Malaysia regards them as refugees and permits them to reside legally, but temporarily, in Sabah. Another 450,000 are living in refugee-like conditions in Malaysia.

*Source: US Committee for Refugees*

The total number of refugees and asylum seekers increased by about 600,000 in 2001, reversing a decade-long downward trend for a third year, according to USCR. That downward trend was due to many factors, including refugee repatriations to several countries and to the unwillingness of many states, especially those in the developed world, to accept new refugees and asylum seekers. The number of refugees and asylum seekers increased in the Middle East, south and central Asia, east Asia and the Americas during the year, largely because of ongoing or renewed conflicts. More than half of the world's refugees and asylum seekers were found in just four countries or territories: Iran, Jordan, the Gaza Strip and the West Bank, and Pakistan.

## Table 16 Significant populations of internally displaced people (1992 to 2001)

| | 1995 | 1996 | 1997 | 1998 | 1999 | 2000 | 2001 |
|---|---|---|---|---|---|---|---|
| **Africa** | **10,185,000** | **8,805,000** | **7,590,000** | **8,958,000** | **10,355,000** | **10,527,000** | **10,835,000** |
| Algeria | – | 10,000 | n.a. | 200,000 | 100,000 | 100,000 | 100,000 |
| Angola | 1,500,000 | 1,200,000 | 1,200,000 | 1,500,000 | 1,500,000 | 2,000,000 | 2,000,000 |
| Burundi | 300,000 | 400,000 | 500,000 | 500,000 | 800,000 | 600,000 | 600,000 |
| Central African Republic | – | – | – | – | – | – | 5,000 |
| Congo, DR[1] | 225,000 | 400,000 | 100,000 | 300,000 | 800,000 | 1,500,000 | 50,000 |
| Congo, PR | – | – | – | 250,000 | 500,000 | 30,000 | 2,000,000 |
| Côte d'Ivoire | – | – | – | – | – | 2,000 | 5,000 |
| Djibouti | – | 25,000 | 5,000 | – | – | – | – |
| Eritrea | – | – | – | 100,000 | 250,000 | 310,000 | 90,000 |
| Ethiopia | – | – | – | 150,000 | 300,000 | 250,000 | 100,000 |
| Ghana | 150,000 | 20,000 | 20,000 | 20,000 | – | – | – |
| Guinea | – | – | – | – | – | 60,000 | 100,000 |
| Guinea-Bissau | – | – | – | 200,000 | 50,000 | – | – |
| Kenya | 210,000 | 100,000 | 150,000 | 200,000 | 100,000 | 100,000 | 200,000 |
| Liberia | 1,000,000 | 1,000,000 | 500,000 | 75,000 | 50,000 | 20,000 | 80,000 |
| Mozambique | 500,000 | – | – | – | – | – | – |
| Nigeria | – | 30,000 | 50,000 | 3,000 | 5,000 | – | 50,000 |
| Rwanda | 500,000 | – | 50,000 | 500,000 | 600,000 | 150,000 | – |
| Senegal | – | – | 10,000 | 10,000 | – | 5,000 | 5,000 |
| Sierra Leone | 1,000,000 | 800,000 | 500,000 | 300,000 | 500,000 | 700,000 | 600,000 |
| Somalia | 300,000 | 250,000 | 200,000 | 250,000 | 350,000 | 300,000 | 400,000 |
| South Africa | 500,000 | 500,000 | 5,000 | – | – | – | – |
| Sudan | 4,000,000 | 4,000,000 | 4,000,000 | 4,000,000 | 4,000,000 | 4,000,000 | 4,000,000 |
| Uganda | – | 70,000 | 300,000 | 400,000 | 450,000 | 400,000 | 400,000 |
| Zimbabwe | – | – | – | – | – | – | 50,000 |
| **East Asia and Pacific** | **555,000** | **1,070,000** | **800,000** | **1,150,000** | **1,577,000** | **1,670,000** | **2,266,000** |
| Cambodia | 55,000 | 32,000 | 30,000 | 22,000 | – | – | – |
| East Timor | – | – | – | – | 300,000 | – | – |
| Indonesia | – | – | – | – | 440,000 | 800,000 | 1,400,000 |
| Korea, DPR of | – | – | – | – | – | 100,000 | 100,000 |
| Myanmar | 500,000 | 1,000,000 | 750,000 | 1,000,000 | 600,000 | 600,000 | 600,000 |
| Papua New Guinea | – | 70,000 | 20,000 | 6,000 | 5,000 | – | 1,000 |
| Philippines | – | – | – | 122,000 | 200,000 | 140,000 | 135,000 |
| Solomon Islands | – | – | – | – | 32,000 | 30,000 | 30,000 |
| **South and central Asia** | **1,600,000** | **2,400,000** | **2,253,500** | **2,130,000** | **1,617,000** | **1,542,000** | **2,469,000** |
| Afghanistan | 500,000 | 1,200,000 | 1,250,000 | 1,000,000 | 500,000 | 375,000 | 1,000,000 |
| Bangladesh | – | – | – | 50,000 | 50,000 | 60,000 | 160,000 |

| | 1995 | 1996 | 1997 | 1998 | 1999 | 2000 | 2001 |
|---|---|---|---|---|---|---|---|
| India | 250,000 | 250,000 | 200,000 | 520,000 | 507,000 | 507,000 | 507,000 |
| Pakistan | – | – | – | – | – | – | 2,000 |
| Sri Lanka | 850,000 | 900,000 | 800,000 | 560,000 | 560,000 | 600,000 | 800,000 |
| Tajikistan | – | 50,000 | 3,500 | – | – | – | – |
| **Middle East** | **1,700,000** | **1,475,000** | **1,475,000** | **1,575,000** | **1,917,000** | **1,700,000** | **1,720,000** |
| Gaza Strip and West Bank | – | – | – | – | 17,000 | – | 20,000 |
| Iraq | 1,000,000 | 900,000 | 900,000 | 1,000,000 | 900,000 | 700,000 | 700,000 |
| Israel | – | – | – | – | 200,000 | 200,000 | 200,000 |
| Lebanon | 400,000 | 450,000 | 450,000 | 450,000 | 350,000 | 350,000 | 300,000 |
| Syria[2] | – | 125,000 | 125,000 | 125,000 | 450,000 | 450,000 | 500,000 |
| Yemen | 300,000 | – | – | – | – | – | – |
| **Europe** | **5,080,000** | **4,735,000** | **3,695,000** | **3,685,000** | **3,993,000** | **3,539,000** | **3,130,000** |
| Armenia | 75,000 | 50,000 | 70,000 | 60,000 | – | – | 50 |
| Azerbaijan | 670,000 | 550,000 | 550,000 | 576,000 | 568,000 | 575,000 | 600,000 |
| Bosnia and Herzegovina | 1,300,000 | 1,000,000 | 800,000 | 836,000 | 830,000 | 518,000 | 556,000 |
| Croatia | 240,000 | 185,000 | 110,000 | 61,000 | 50,000 | 34,000 | 23,000 |
| Cyprus | 265,000 | 265,000 | 265,000 | 265,000 | 265,000 | 265,000 | 265,000 |
| Georgia | 260,000 | 280,000 | 285,000 | 275,000 | 280,000 | 272,000 | 264,000 |
| Macedonia | – | – | – | – | – | – | 21,000 |
| Russian Federation | 250,000 | 400,000 | 375,000 | 350,000 | 800,000 | 800,000 | 474,000 |
| Turkey | 2,000,000 | 2,000,000 | 1,250,000 | 1,000,000 | 600,000 | 600,000 | 600,000 |
| Yugoslavia | – | – | – | 257,000 | 600,000 | 475,000 | 277,000 |
| **Americas and Caribbean** | **1,280,000** | **1,220,000** | **1,624,000** | **1,755,000** | **1,886,000** | **2,176,000** | **2,455,000** |
| Colombia | 600,000 | 600,000 | 1,000,000 | 1,400,000 | 1,800,000 | 2,100,000 | 2,440,000 |
| Guatemala | 200,000 | 200,000 | 250,000 | – | – | – | – |
| Mexico | – | – | 14,000 | 15,000 | 16,000 | 16,000 | 15,000 |
| Peru | 480,000 | 420,000 | 360,000 | 340,000 | 70,000 | 60,000 | – |
| **Total** | **20,400,000** | **19,705,000** | **17,437,500** | **19,253,000** | **21,345,000** | **21,154,000** | **22,875,000** |

Notes: – indicates zero or near zero; n.a. not available, or reported estimates unreliable; [1] formerly Zaire; [2] includes about 125,000 IDPs originally displaced from the Golan Heights in 1967 and their progeny.

*Source: US Committee for Refugees*

More than 1.7 million more people were internally displaced at the end of 2001 than at the end of 2000, a year in which the number of internally displaced people also increased. Large internally displaced populations remained in Angola, Azerbaijan, Bosnia and Herzegovina, Burundi, India, Iraq, Myanmar, Russian Federation, Sierra Leone, Sudan, Syria, Turkey and elsewhere in 2001. Significant new displacement occurred in Afghanistan, Central African Republic, Colombia, Democratic Republic of the Congo, Guinea, Indonesia, Liberia, Nigeria and Sri Lanka, while some internally displaced people were able to return home in Eritrea, Ethiopia, Uganda and Yugoslavia. About half of the world's internally displaced people were in Africa. It is important to note that estimates of the size of internally displaced populations are frequently subject to great margins of error and are often imprecise.

Section Two
**Tracking
the system**

# International Federation overview

In 2001, the International Federation continued to focus its work on activities in the four core areas defined in *Strategy 2010*: disaster response; disaster preparedness; health and care in the community; and promotion of the International Red Cross and Red Crescent Movement's Fundamental Principles and humanitarian values.

There were 711 disaster events in 2001 and the International Federation, through its regional and country delegations and in collaboration with National Red Cross and Red Crescent Societies, was on hand to respond to many of them. Some, such as the earthquakes in India and El Salvador, received much media coverage; many others were hardly noticed by the outside but wrought extensive disruption on the lives and livelihoods of the vulnerable people affected.

During the year, the International Federation took part in a number of health initiatives, including vaccination campaigns against measles and polio, in collaboration with a number of other organizations. It was also active in encouraging greater openness and more engagement in the fight against HIV/AIDS.

The United Nations (UN) proclaimed 2001 as International Year of Volunteers (IYV2001). As the largest volunteer organization in the world, the International Federation took an active part in IVY2001, helping national Red Cross and Red Crescent societies to improve their work of recruiting, selecting, training, organizing and rewarding volunteers, and to promote the volunteering environment in their countries.

From 1990 to 2000, the number of beneficiaries targeted for International Federation support was ever increasing: from 5 million in 1990, 30 million in 1999 and 50 million in 2000. In 2001, the number of people in need of assistance as a result of floods, droughts, earthquakes displacement and health emergencies dropped to 34 million. To assist these people in need, the International Federation launched relief appeals for 480 million Swiss francs, of which 343 million Swiss francs was sought for the annual appeal covering long-term operations and 137 million francs was requested for emergencies. The biggest emergency appeal of the year, for some 28.7 million francs, was in response to the humanitarian crisis in Afghanistan.

The International Federation remains committed to improving the quality of humanitarian assistance and accountability. In 1994, along with seven other humanitarian agencies (Caritas, Internationalis, Catholic Relief Services, International Committee

Photo opposite page: Working with national Red Cross and Red Crescent societies, the International Federation remains committed to improving the lives of the most vulnerable people around the world.

Howard Davies/Exile Images, Nepal 1997.

## Box 9.1 IDRL development continues

Laws requiring and guiding rapid and efficient cross-border relief can save lives.

Consultations with governments, national Red Cross and Red Crescent societies and other experts have made clear a widespread wish to establish coherent links between the various pieces of "hard" and "soft" law that exist to guide international relief for natural disasters, as well as to identify gaps and the best ways of filling them. The International Federation maintains its leadership role in the development of this major initiative. Governments and international organizations, as well as academics and field practitioners, see a body of international disaster response law (IDRL) as having the capacity to ensure far more effective humanitarian assistance and protection to affected communities, who suffer from the consequences of natural and technological disasters, and improved efficiency in the delivery.

Since the first presentation of IDRL in the World Disasters Report 2000 and the update in the 2001 report, the International Red Cross and Red Crescent Movement's Council of Delegates took up the issue at its session in November 2001. In an important resolution it endorsed the International Federation's initiative to advocate for the development and, where applicable, the improvement and faithful application of IDRL. The work at the international level will concentrate on, but not be limited to, the compilation and publication of existing international laws and regulations, and the evaluation of their actual effectiveness in humanitarian operations.

Work done since then has included an important study, commissioned by the International Federation's secretariat, which compiles existing law.

First outcomes of the study clearly show that a considerable body of law exists in one form or another. It is also evident that different law exists in different regions of the world. An important element in International Federation planning is therefore to involve comprehensively in the project, governments and experts, including universities, which represent regional perspectives. The plan is to complete an analysis of existing law and publish a first compilation of legal texts at the end of 2002. A first practical step could be to produce a compendium in a form that could be used for easy reference in field situations. But coherence is absolutely necessary and there appear to be many gaps.

Parallel to, and complementing, this legal study, the International Federation is seeking support from a number of governments and National Societies to undertake field studies to gain a better-documented understanding of the problems practitioners can face that could be addressed through the mechanism of IDRL. This will provide a sound basis for further consideration of options for action.

The International Federation plan envisages detailed work on options during 2003, involving governments, National Societies and experts. The outcome should be a full report presented to the International Red Cross and Red Crescent Conference in December 2003, with recommendations for work beyond 2003. ∎

of the Red Cross, International Save the Children Alliance, Lutheran World Federation, Oxfam and World Council of Churches), it developed the *Code of Conduct for the International Red Cross and Red Crescent Movement and Non-*

*Governmental Organizations in Disaster Relief.* The voluntary code sets out universal basic standards to govern the way signatory relief agencies should work in disaster assistance. As of 31 March 2002, 198 non-governmental organizations (NGOs) had become signatories to the code, and had agreed to incorporate the code's ten points of principle into their work. For more information, visit the *Code of conduct* page on the International Federation's web site at http://www.ifrc.org/publicat/conduct/

The International Federation continued to support a number of projects including:
- Sphere, an international interagency effort which provides humanitarian agencies with a framework for rights-based humanitarian assistance and adherence to minimum standards. More information about the Sphere project can be found at http://www.sphereproject.org/
- The Global Road Safety Partnership (GRSP), whose mission is to increase the safety on the roads in developing countries using a partnership approach where stakeholders from government, civil society and business work together and contribute, in the most appropriate way for each organization, to the improvement of road safety. http://www.grsproadsafety.org/
- Reach Out, a joint initiative of several international humanitarian agencies on refugee protection training. A three-year programme, it aims to disseminate basic refugee protection knowledge among members of NGOs and the Red Cross Red Crescent. http://www.reachout.ch/index.htm

The International Federation is also a member of the task force of the International Strategy for Disaster Reduction (ISDR), which was adopted in 2000 by the UN General Assembly as a framework for action in support of disaster reduction activities. To support its actions, ISDR will publish a *Global Review of Disaster Reduction* in 2002. The review presents indications of global trends in disaster risk reduction practices, through an overview of disaster reduction-related initiatives, programmes and institutional structures. ISDR intends to initiate a structured process to maintain and disseminate periodically information on disaster reduction. For more information, visit the ISDR web site (http://www.unisdr.org).

The International Federation continued to follow up on an issue raised in the *World Disasters Report 2000* on the necessity for an international disaster response law (see Box 9.1).

chapter 10

Section Two

**Tracking
the system**

National Societies:
mobilizing the
power of humanity
to assist vulnerable
people around
the world.

Ecuadorian
Red Cross,
Ecuador 2001.

# The Movement:
# a worldwide presence

Contact details for the members of the International Red Cross and Red Crescent Movement

## International Federation of Red Cross and Red Crescent Societies
P.O. Box 372
1211 Geneva 19
Switzerland
Tel. (41)(22) 7304222
Fax (41)(22) 7330395
Tlx (045) 412 133 FRC CH
Tlg. LICROSS GENEVA
E-mail secretariat@ifrc.org
Web http://www.ifrc.org

## International Committee of the Red Cross
19 avenue de la Paix
1202 Geneva
Switzerland
Tel. (41)(22) 734 60 01
Fax (41)(22) 733 20 57
Tlx 414 226 CCR CH
Tlg. INTERCROIXROUGE GENEVE
E-mail icrc.gva@gwn.icrc.org
Web http://www.icrc.org

## National Red Cross and Red Crescent Societies

National Red Cross and Red Crescent Societies are listed alphabetically by International Organization for Standardization Codes for the Representation of Names of Countries, English spelling.

Details correct as of 1 March 2002. Please forward any corrections to the International Federation's Information Resource Centre in Geneva (e-mail: irc@ifrc.org).

### Afghan Red Crescent Society
Pul Artel
P.O. Box 3066
Shar-e-Now
Kabul
Afghanistan
Tel. (873) (628) 32537

### Albanian Red Cross
Rruga "Muhammet Gjollesha"
Sheshi "Karl Topia"
C.P. 1511
Tirana
Albania
Tel. (355)(42) 25855
Fax (355)(42) 22037

### Algerian Red Crescent
15 bis, Boulevard Mohammed V
Alger 16000
Algeria
Tel. (213) (21) 633956
Fax (213) (21) / 633690
Tlx 56056 HILAL ALGER
E-mail cra@algeriainfo.com

## Andorra Red Cross

Prat de la Creu 22
Andorra la Vella
Andorra
Tel. (376) 825225
Fax (376) 828630
E-mail creuroja@creuroja.ad
Web http://www.creuroja.ad

## Angola Red Cross

Rua 1° Congresso no 21
Caixa Postal 927
Luanda
Angola
Tel. (244)(2) 336543
Fax (244)(2) 345065
Tlx 3394 CRUZVER AN

## Antigua and Barbuda Red Cross Society

Red Cross Headquarters
Old Parham Road
P.O. Box 727
St. Johns, Antigua W.I.
Antigua and Barbuda
Tel. (1)(268) 4620800
Fax (1)(268) 4609595
E-mail redcross@candw.ag

## Argentine Red Cross

Hipólito Yrigoyen 2068
1089 Buenos Aires
Argentina
Tel. (54)(114) 9511391
Fax (54)(114) 9527715
Tlx 21061 CROJA AR
Web http://www.cruzroja.org.ar

## Armenian Red Cross Society

21 Paronian Street
375015 Yerevan
Armenia
Tel. (374)(2) 538064
Fax (374)(2) 151129
E-mail redcross@redcross.am

## Australian Red Cross

155 Pelham Street
P.O. Box 196
Carlton South VIC 3053
Australia
Tel. (61)(3) 93451800
Fax (61)(3) 93482513
E-mail redcross@nat.redcross.org.au
Web http://www.redcross.org.au

## Austrian Red Cross

Wiedner Hauptstrasse 32
Postfach 39
1041 Wien
Austria
Tel. (43)(1) 58900 0
Fax (43)(1) 58900 199
E-mail oerk@redcross.or.at
Web http://www.roteskreuz.at

## Red Crescent Society of Azerbaijan

Prospekt Azerbaidjan 19
Baku
Azerbaijan
Tel. (994)(12) 938481
Fax (994)(12) 931578
E-mail azerb.redcrescent@azdata.net

## The Bahamas Red Cross Society

P.O. Box N-8331
Nassau
Bahamas
Tel. (1)(242) 3237370
Fax (1)(242) 3237404
E-mail redcross@bahamas.net.bs
Web http://www.bahamasrc.com

## Bahrain Red Crescent Society

P.O. Box 882
Manama
Bahrain
Tel. (973) 293171
Fax (973) 291797
E-mail hilal@baletco.com.bh

## Bangladesh Red Crescent Society

684-686 Bara Maghbazar
G.P.O. Box 579
Dhaka - 1217
Bangladesh
Tel. (880)(2) 9330188
Fax (880)(2) 9352303
E-mail bdrcs@bdonline.com

## The Barbados Red Cross Society

Red Cross House
Jemmotts Lane
Bridgetown
Barbados
Tel. (1)(246) 4262052
Fax (1)(246) 4262052
E-mail bdosredcross@caribsurf.com

## Belarusian Red Cross

35, Karl Marx Str.
220030 Minsk
Belarus
Tel. (375)(17) 2272620
Fax (375)(17) 2272620
E-mail brc@home.by

## Belgian Red Cross

Ch. de Vleurgat 98
1050 Bruxelles
Belgium
Tel. (32)(2) 6454545
Fax (32)(2) 6460439 (French);
    6433406 (Flemish)
E-mail info@redcross-fr.be
    documentatie@redcross-fl.be
Web http://www.redcross.be

## Belize Red Cross Society

1 Gabourel Lane
P.O. Box 413
Belize City
Belize
Tel. (501)(2) 73319
Fax (501)(2) 30998
E-mail bzercshq@btl.net

## Red Cross of Benin

B.P. No. 1
Porto-Novo
Benin
Tel. (229) 212886
Fax (229) 214927
Tlx 1131 CRBEN

## Bolivian Red Cross

Avenida Simón Bolívar N° 1515
Casilla No. 741
La Paz
Bolivia
Tel. (591)(2) 202930
Fax (591)(2) 359102
E-mail cruzrobo@caoba.entelnet
Web http://www.
      cruzrojaboliviana.org

## The Red Cross Society of Bosnia and Herzegovina

Titova 7
71000 Sarajevo
Bosnia and Herzegovina
Tel. (387) (71) 666009
Fax (387) (71) 666010
E-mail rcsbihhq@bih.net.ba

## Botswana Red Cross Society

135 Independence Avenue
P.O. Box 485
Gaborone
Botswana
Tel. (267) 352465
Fax (267) 312352
E-mail brcs@info.bw

## Brazilian Red Cross

Praça Cruz Vermelha No. 10/12
20230-130 Rio de Janeiro RJ
Brazil
Tel. (55)(21) 22210658
Fax (55)(21) 22210658
Tlx (38) 2130532 CVBR BR

## Brunei Darussalam Red Crescent Society

P.O. Box 1315
Kuala Belait KA 1131
Negara
Brunei Darussalam
Tel. (673)(2) 339774
Fax (673)(2) 335314
E-mail bdrcs@netkad.com.bn

## Bulgarian Red Cross

76, James Boucher Boulevard
1407 Sofia
Bulgaria
Tel. (359)(2) 650595
Fax (359)(2) 656937
E-mail secretariat@redcross.bg
Web http://www.redcross.bg

## Burkinabe Red Cross Society

01 B.P. 4404
Ouagadougou 01
Burkina Faso
Tel. (226) 361340
Fax (226) 363121
Tlx LSCR 5438 BF
      OUAGADOUGOU

## Burundi Red Cross

18, Av. de la Croix-Rouge
B.P. 324
Bujumbura
Burundi
Tel. (257) 216246
Fax (257) 211101
E-mail croixrou@cbinf.com

## Cambodian Red Cross Society

17, Vithei de la Croix-Rouge
Cambodgienne (180)
Phnom-Penh
Cambodia
Tel. (855)(23) 212876
Fax (855)(23) 212875
E-mail crc@camnet.com.kh

## Cameroon Red Cross Society

Rue Henri Dunant
B.P. 631
Yaoundé
Cameroon
Tel. (237) 3224177
Fax (237) 3224177
Tlx (0970) 8884 KN

## The Canadian Red Cross Society

170 Metcalfe Street, Suite 300
Ottawa
Ontario K2P 2P2
CANADA
Tel. (1)(613) 7401900
Fax (1)(613) 7401911
E-mail cancross@redcross.ca
Web http://www.redcross.ca

## Red Cross of Cape Verde

Rua Andrade Corvo
Caixa Postal 119
Praia
Cape Verde
Tel. (238) 611701
Fax (238) 614174
Tlx 6004 CV CV

## Central African Red Cross Society

Avenue Koudoukou, Km. 5
B.P. 1428
Bangui
Central African Republic
Tel. (236) 612223
Fax (236) 613561
Tlx DIPLOMA 5213

## Red Cross of Chad

B.P. 449
N'Djamena
Chad
Tel. (235) 523434
Fax (235) 525218
E-mail croix-rouge@intnet.td

## Chilean Red Cross

Avenida Santa María No. 150
Providencia
Correo 21, Casilla 246 V
Santiago de Chile
Chile
Tel. (56)(2) 7771448
Fax (56)(2) 7370270
E-mail cruzroja@rdc.cl

## Red Cross Society of China

53 Ganmian Hutong
100010 Beijing
China
Tel. (86)(10) 65124447
Fax (86)(10) 65124169
E-mail rcsc@chineseredcross.org.cn
Web http://www.
    chineseredcross.org.cn

## Colombian Red Cross Society

Avenida 68 N° 66-31
Apartado Aéreo 1110
Bogotá D.C.
Colombia
Tel. (57)(1) 4376339/ 4289423
Fax (57)(1) 4281725 / 4376301
E-mail inter@andinet.com
Web http://www.crcol.org.co/

## Congolese Red Cross

Place de la Paix
B.P. 4145
Brazzaville
Congo
Tel. (242) 824410
Fax (242) 828825
E-mail: sjm@ficr.aton.cd

## Red Cross of the Democratic Republic of the Congo

41, Avenue de la Justice
Zone de la Gombe
B.P. 1712
Kinshasa I
Congo, D.R. of the
Tel. (243)(12) 34897
Fax (243) 8804551
E-mail: secretariat@crrdc.aton.cd

## Costa Rican Red Cross

Calle 14, Avenida 8
Apartado 1025
San José 1000
Costa Rica
Tel. (506) 2337033
Fax (506) 2237628
E-mail por info@cruzroja.or.cr
Web http://www.cruzroja.or.cr

## Red Cross Society of Côte d'Ivoire

P.O. Box 1244
Abidjan 01
Côte d'Ivoire
Tel. (225) 22321335
Fax (225) 22225355
Tlx 24122 SICOGI CI

## Croatian Red Cross

Ulica Crvenog kriza 14
10000 Zagreb
Croatia
Tel. (385)(1) 4655814
Fax (385)(1) 4655365
E-mail redcross@hck.hr
Web http://www.hck.hr/

## Cuban Red Cross

Calle 20#707
Miramar Playa
C.D. 11300 Cuba
Cuidad de la Habana
Cuba
Tel. (53)(7) 225591
Fax (53)(7) 228272
E-mail crsn@infomed.sld.cu

## Czech Red Cross

Thunovska 18
CZ-118 04 Praha 1
Czech Republic
Tel. (420)(2) 51104111
Fax (420)(2) 51104261
E-mail cck.zahranicni@iol.cz

## Danish Red Cross

Blegdamsvej 27
P.O. Box 2600
DK-2100 København Ø
Denmark
Tel. (45) 35259200
Fax (45) 35259292
E-mail drc@redcross.dk
Web http://www.redcross.dk

## Red Crescent Society of Djibouti

B.P. 8
Djibouti
Djibouti
Tel. (253) 352451
Fax (253) 351505
Tlx 5871 PRESIDENCE DJ

## Dominica Red Cross Society

Federation Drive
Goodwill
Dominica
Tel. (1)(767) 4488280
Fax (1)(767) 4487708
E-mail redcross@cwdom.dm

## Dominican Red Cross

Calle Juan E. Dunant No. 51
Ens. Miraflores
Apartado Postal 1293
Santo Domingo, D.N.
Dominican Republic
Tel. (1)(809) 6823793
Fax (1)(809) 6822837
E-mail cruz.roja@codetel.net.do

## Ecuadorian Red Cross
Antonio Elizalde E 4-31
y Av. Colombia (esq.)
Casilla 1701 2119
Quito
Ecuador
Tel. (593)(2) 570424
Fax (593)(2) 570424
E-mail difusio@attglobal.net

## Egyptian Red Crescent Society
29, El Galaa Street
Cairo
Egypt
Tel. (20)(2) 5750558
Fax (20)(2) 5740450
E-mail erc@brainy1.ie-eg.com

## Salvadorean Red Cross Society
17 C. Pte. y Av. Henri Dunant
Apartado Postal 2672
San Salvador
El Salvador
Tel. (503) 2227749
Fax (503) 2227758
E-mail crsalvador@vianet.com.sv

## Red Cross of Equatorial Guinea
Alcalde Albilio Balboa 92
Apartado postal 460
Malabo
Equatorial Guinea
Tel. (240)(9) 3701
Fax (240)(9) 3701

## Estonia Red Cross
Lai Street 17
EE0001 Tallinn
Estonia
Tel. (372) 6411643
Fax (372) 6411641
E-mail haide.laanemets@recross.ee

## Ethiopian Red Cross Society
Ras Desta Damtew Avenue
P.O. Box 195
Addis Ababa
Ethiopia
Tel. (251)(1) 519364
Fax (251)(1) 512643
E-mail ercs@telecom.net.et

## Fiji Red Cross Society
22 Gorrie Street
GPO Box 569
Suva
Fiji
Tel. (679) 314133
Fax (679) 303818
E-mail redcross@is.com.fj

## Finnish Red Cross
Tehtaankatu 1 a
P.O. Box 168
FIN-00141 Helsinki
Finland
Tel. (358)(9) 12931
Fax (358)(9) 1293326
E-mail forename.surname@redcross.fi
Web http://www.redcross.fi

## French Red Cross
1, Place Henry-Dunant
F-75384 Paris Cedex 08
France
Tel. (33)(1) 44431100
Fax (33)(1) 44431101
E-mail contact@croix-rouge.net
Web http://www.croix-rouge.fr

## Gabonese Red Cross Society
Place de l'Indépendance
Derrière le Mont de Cristal
Boîte Postale 2274
Libreville
Gabon
Tel. (241) 766159
Fax (241) 766160
E-mail gab.cross@internetgabon.com

## The Gambia Red Cross Society
Kanifing Industrial Area - Banjul
P.O. Box 472
Banjul
Gambia
Tel. (220) 392405
Fax (220) 394921
E-mail redcrossgam@delphi.com

## Red Cross Society of Georgia
15, Krilov St.
38002 Tbilisi
Georgia
Tel. (995)(32)961534
Fax (995)(32) 953304
E-mail grc@caucasus.net

## German Red Cross
Carstennstrasse 58
D-12205 Berlin
Germany
Tel. (49) (30) 85404-0
Fax (49) (30) 85404470
E-mail drk@drk.de
Web http://www.rotkreuz.de

## Ghana Red Cross Society
Ministries Annex Block A3
Off Liberia Road Extension
P.O. Box 835
Accra
Ghana
Tel. (233)(21) 662298
Fax (233)(21) 667226
E-mail grcs@idngh.com

## Hellenic Red Cross
Rue Lycavittou 1
Athens 106 72
Greece
Tel. (30)(10) 3250515
Fax (30)(10) 3250577
E-mail ir@redcross.gr
Web http://www.redcross.gr

## Grenada Red Cross Society

Upper Lucas Street
P.O. Box 551
St. George's
Grenada
Tel. (1)(473) 4401483
Fax (1)(473) 4401829
E-mail grercs@caribsurf.com

## Guatemalan Red Cross

3a Calle 8-40, Zona 1
Guatemala, C.A.
Guatemala
Tel. (502)(2) 322026
Fax (502)(2) 324649
E-mail crg@guate.net

## Red Cross Society of Guinea

B.P. 376
Conakry
Guinea
Tel. (224) 412310
Fax (224) 414255
Tlx 22101

## Red Cross Society of Guinea-Bissau

Avenida Unidade Africana, No. 12
Caixa postal 514-1036 BIX, Codex
Bissau
Guinea-Bissau
Tel. (245) 202408
Tlx 251 PCE BI

## The Guyana Red Cross Society

Eve Leary
P.O. Box 10524
Georgetown
Guyana
Tel. (592)(2) 65174
Fax (592)(2) 77099
E-mail redcross@sdnp.org.gy
Web http://www.
    sdnp.org.gy/redcross/

## Haitian National Red Cross Society

1, rue Eden
Bicentenaire
CRH, B.P. 1337
Port-Au-Prince
Haiti
Tel. (509) 5109813
Fax (509) 2231054
E-mail croroha@publi-tronic.com

## Honduran Red Cross

7a Calle
entre 1a. y 2a. Avenidas
Comayagüela D.C.
Honduras
Tel. (504) 2378876
Fax (504) 2380185
E-mail honducruz@datum.hn

## Hungarian Red Cross

Arany János utca 31
Magyar Vöröskereszt
1367 Budapest 5, Pf. 121
Hungary
Tel. (36)(1) 3741338
Fax (36)(1) 3741312
E-mail intdept@hrc.hu

## Icelandic Red Cross

Efstaleiti 9
103 Reykjavik
Iceland
Tel. (354) 5704000
Fax (354) 5704010
E-mail central@redcross.is
Web http://www.redcross.is/

## Indian Red Cross Society

Red Cross Building
1 Red Cross Road
New Delhi 110001
India
Tel. (91)(11) 3716441
Fax (91)(11) 3717454
E-mail indcross@vsnl.com

## Indonesian Red Cross Society

Jl. Jenderal Datot Subroto Kav. 96
P.O. Box 2009
Jakarta 12790
Indonesia
Tel. (62)(21) 7992325
Fax (62)(21) 7995188
Tlx 66170 MB PMI IA

## Red Crescent Society of the Islamic Republic of Iran

Ostad Nejatolahi Ave.
Tehran
Iran, Islamic Republic of
Tel. (98)(21) 8849077
Fax (98)(21) 8849079
E-mail helal@mail.dci.co.ir
Web http://www.irrcs.org

## Iraqi Red Crescent Society

Al-Mansour
P.O. Box 6143
Baghdad
Iraq
Tel. (964)(1) 8862191
Fax (964)(1) 5372519
Tlx 213331 HELAL IK

## Irish Red Cross Society

16, Merrion Square
Dublin 2
Ireland
Tel. (353)(1) 6765135
Fax (353)(1) 6614461
E-mail redcross@iol.ie
Web http://www.redcross.ie

## Italian Red Cross

Via Toscana 12
I-00187 Roma - RM
Italy
Tel. (39)(06) 47591
Fax (39)(06) 4759223
Web http://www.cri.it/

## Jamaica Red Cross

Central Village
Spanish Town, St. Catherine
76 Arnold Road
Kingston 5
Jamaica
Tel. (1)(876) 98478602
Fax (1)(876) 9848272
E-mail jrcs@infochan.com
Web http://www.
  infochan.com/ja-red-cross/

## Japanese Red Cross Society

1-3 Shiba Daimon, 1-Chome,
Minato-ku
Tokyo-105-8521
Japan
Tel. (81)(3) 34377087
Fax (81)(3) 34358509
E-mail kokusai@jrc.or.jp
Web http://www.jrc.or.jp

## Jordan National Red Crescent Society

Madaba Street
P.O. Box 10001
Amman 11151
Jordan
Tel. (962)(64) 773141
Fax (962)(64) 750815
E-mail jrc@index.com.jo

## Kenya Red Cross Society

Nairobi South "C"
(Belle Vue), off Mombasa Road
P.O. Box 40712
Nairobi
Kenya
Tel. (254)(2) 603593
Fax (254)(2) 603589
Tlx. 25436 IFRC KE

## Kiribati Red Cross Society

P.O. Box 213
Bikenibeu
Tarawa
Kiribati
Tel. (686) 28128
Fax (686) 21416
E-mail redcross@tskl.net.ki

## Red Cross Society of the Democratic People's Republic of Korea

Ryonwa 1, Central District
Pyongyang
Korea, Democratic People's
Republic of
Tel. (850)(2) 18333
Fax (850)(2) 3814644
Tlg. KOREACROSS
  PYONGYANG

## The Republic of Korea National Red Cross

32 - 3ka, Namsan-dong
Choong-Ku
Seoul 100 - 043
Korea, Republic of
Tel. (82)(2) 37053705
Fax (82)(2) 37053667
E-mail knrc@redcross.or.kr
Web http://www.redcross.or.kr

## Kuwait Red Crescent Society

Al-Jahra St.
Shuweek
P.O. Box 1359
13014 Safat
Kuwait
Tel. (965) 4815478
Fax (965) 4839114
E-mail krcs@kuwait.net

## Red Crescent Society of Kyrgyzstan

10, prospekt Erkindik
720040 Bishkek
Kyrgyzstan
Tel. (996)(312) 222414
Fax (996)(312) 662181
E-mail redcross@imfiko.bishkek.su

## Lao Red Cross

Avenue Sethathirath
B.P. 650
Vientiane
Lao People's Democratic Republic
Tel. (856)(21) 222398
Fax (856)(21) 212128
Tlx 4491 TE via PTT LAOS

## Latvian Red Cross

1, Skolas Street
Riga, LV-1010
Latvia
Tel. (371) 7336650
Fax (371) 7336651
E-mail secretariat@redcross.lv

## Lebanese Red Cross

Rue Spears
Beyrouth
Lebanon
Tel. (961)(1) 372802
Fax (961)(1) 378207
E-mail redcross@dm.net.lb
Web http://www.dm.net.lb/redcross/

## Lesotho Red Cross Society

23 Mabile Road
P.O. Box 366
Maseru 100
Lesotho
Tel. (266) 313911
Fax (266) 310166
E-mail lesoff@lesred.co.za

## Liberian Red Cross Society

107 Lynch Street
P.O. Box 20-5081
1000 Monrovia 20
Liberia
Tel. 888-330125
Fax (231) 330125
E-mail lnrc@Liberia.net

## Libyan Red Crescent

P.O. Box 541
Benghazi
Libyan Arab Jamahiriya
Tel. (218)(61) 9095202
Fax (218)(61) 9095829
E-mail libyan_redcrescent@
  libyamail.net

## Liechtenstein Red Cross

Heiligkreuz 25
FL-9490 Vaduz
Liechtenstein
Tel. (423) 2322294
Fax (423) 2322240
E-mail info@lieredcross.li

## Lithuanian Red Cross Society

Gedimino ave. 3a
2600 Vilnius
Lithuania
Tel. (370)(2) 628037
Fax (370)(2) 619923
E-mail redcross@tdd.lt
Web http://www.redcross.lt

## Luxembourg Red Cross

Parc de la Ville
B.P. 404
L - 2014 Luxembourg
Tel. (352) 450202
Fax (352) 457269
E-mail siege@croix-rouge.lu
Web http://www.croix-rouge.lu/

## The Red Cross of The Former Yugoslav Republic of Macedonia

No. 13
Bul. Koco Racin
91000 Skopje
Macedonia, The Former
    Yugoslav Republic of
Tel. (389)(91) 114355
Fax (389)(91) 230542

## Malagasy Red Cross Society

1, rue Patrice Lumumba
Tsavalalana
B.P. 1168
Antananarivo
Madagascar
Tel. (261)(20) 2222111
Fax (261)(20) 2235457
E-mail crm@dts.mg

## Malawi Red Cross Society

Red Cross House (along
Presidential Way)
P.O. Box 30096
Capital City, Lilongwe 3
Malawi
Tel. (265) 775291
Fax (265) 775590
E-mail mrcs@eomw.net

## Malaysian Red Crescent Society

JKR 32, Jalan Nipah
Off Jalan Ampang
55000 Kuala Lumpur
Malaysia
Tel. (60)(3) 42578122
Fax (60)(3) 4533191
E-mail mrcs@po.jaring.my
Web http://www.redcrescent.org.my/

## Mali Red Cross

Route Koulikoro
B.P. 280
Bamako
Mali
Tel. (223) 224569
Fax (223) 240414
Tlx 2611 MJ

## Malta Red Cross Society

104 St Ursula Street
Valletta VLT 05
Malta
Tel. (356) 222645
Fax (356) 243664
E-mail redcross@waldonet.net.mt
Web http://www.redcross.org.mt/

## Mauritanian Red Crescent

Avenue Gamal Abdel Nasser
B.P. 344
Nouakchott
Mauritania
Tel. (222) 6307510
Fax (222) 5291221
Tlx 5830 CRM

## Mauritius Red Cross Society

Ste. Thérèse Street
Curepipe
Mauritius
Tel. (230) 6763604
Fax (230) 6748855
E-mail redcross@intnet.mu

## Mexican Red Cross

Calle Luis Vives 200
Colonia Polanco
México, D.F. 11510
Mexico
Tel. (52)(5) 3950606
Fax (52)(5) 3951598
E-mail cruzroja@mexporta.com
Web http://www.cruz-roja.org.mx/

## Red Cross Society of the Republic of Moldova

67a, Ulitsa Asachi
MD-277028 Chisinau
Moldova
Tel. (373)(2) 729644
Fax (373)(2) 729700
E-mail moldova-RC@mdl.net

## Red Cross of Monaco

27, Boulevard de Suisse
Monte Carlo
Monaco
Tel. (377)(97) 976800
Fax (377)(93) 159047
E-mail redcross@monaco.mc
Web http://www.croix-rouge.mc

## Mongolian Red Cross Society

Central Post Office
Post Box 537
Ulaanbaatar 13
Mongolia
Tel. (976) (1) 321864
Fax (976) (1) 321864
E-mail redcross@magicnet.mn

## Moroccan Red Crescent

Palais Mokri
Takaddoum
B.P. 189
Rabat
Morocco
Tel. (212)(7) 650898
Fax (212)(7) 759395
E-mail crm@iam.net.ma

## Mozambique Red Cross Society

Avenida Agostinho Neto 284
Caixa Postal 2488
Maputo
Mozambique
Tel. (258)(1) 490943
Fax (258)(1) 497725
E-mail cvm@mail.tropical.co.mz

## Myanmar Red Cross Society

42 Strand Road
Yangon
Myanmar
Tel. (95)(1) 296552
Fax (95)(1) 296551
Tlx 21218 BRCROS BM

## Namibia Red Cross

Erf 2128, Independence Avenue
Katutura
P.O. Box 346
Windhoek
Namibia
Tel. (264)(61) 235216
Fax (264)(61) 228949
E-mail namcross@iafrica.com.na

## Nepal Red Cross Society

Red Cross Marg
Kalimati
P.O. Box 217
Kathmandu
Nepal
Tel. (977)(1) 270650
Fax (977)(1) 271915
E-mail nrcs@nhqs.wlink.com.np

## The Netherlands Red Cross

Leeghwaterplein 27
P.O. Box 28120
2502 KC The Hague
Netherlands
Tel. (31)(70) 4455666
Fax (31)(70) 4455777
E-mail hq@redcross.nl
Web http://www.rodekruis.nl

## New Zealand Red Cross

69 Molesworth Street
P.O. Box 12-140
Thorndon
Wellington 6038
New Zealand
Tel. (64)(4) 4723750
Fax (64)(4) 4730315
E-mail jcs@redcross.org.nz
Web http://www.redcross.org.nz/

## Nicaraguan Red Cross

Reparto Belmonte
Carretera Sur, km 7
Apartado 3279
Managua
Nicaragua
Tel. (505)(2) 651307
Fax (505)(2) 651643
E-mail nicacruz@ibw.com.ni

## Red Cross Society of Niger

B.P. 11386
Niamey
Niger
Tel. (227) 733037
Fax (227) 732461
E-mail crniger@intnet.ne

## Nigerian Red Cross Society

11, Eko Akete Close
off St. Gregory's Road
South West Ikoyi
P.O. Box 764
Lagos
Nigeria
Tel. (234)(1) 2695188
Fax (234)(1) 2691599
E-mail nrcs@nigerianredcross.org

## Norwegian Red Cross

Hausmannsgate 7
Postbox 1. Gronland
0133 Oslo
Norway
Tel. (47) 22054000
Fax (47) 22054040
E-mail documentation.
    center@redcross.no
Web http://www.redcross.no/

## Pakistan Red Crescent Society

Sector H-8
Islamabad
Pakistan
Tel. (92)(51) 9257404
Fax (92)(51) 9257408
E-mail hilal@comsats.net.pk
Web http://www.prcs.org.pk/

## Palau Red Cross Society

P.O. Box 6043
Koror
Republic of Palau 96940
Tel. (680) 4885780
Fax (680) 4884540
E-mail palredcross@palaunet.com

## Red Cross Society of Panama

Albrook, Areas Revertidas
Calle Principal
Edificio # 453
Apartado 668
Zona 1 Panamá
Panama
Tel. (507) 3151389
Fax (507) 3151401
E-mail cruzroja@pan.gbm.net

## Papua New Guinea Red Cross Society

Taurama Road
Port Moresby
P.O. Box 6545
Boroko
Papua New Guinea
Tel. (675) 3258577
Fax (675) 3259714

## Paraguayan Red Cross

Brasil 216 esq. José Berges
Asunción
Paraguay
Tel. (595)(21) 222797
Fax (595)(21) 211560
E-mail cruzroja@pla.net.py

## Peruvian Red Cross

Av. Arequipa No. 1285
Lima
Peru
Tel. (51)(1) 4710701
Fax (51)(1) 4710701
E-mail cruzrojaperuana@
    terra.com.pe

## The Philippine National Red Cross

Bonifacio Drive
Port Area
P.O. Box 280
Manila 2803
Philippines
Tel. (63)(2) 5270866
Fax (63)(2) 5270857
E-mail secgen_pnrc@email.com

## Polish Red Cross

Mokotowska 14
P.O. Box 47
00-950 Warsaw
Poland
Tel. (48)(22) 6285201
Fax (48)(22) 6284168
E-mail pck@atomnet.pl
Web http://www.pck.org.pl/

## Portuguese Red Cross

Campo Grande, 28-6th
1700-093 Lisboa
Portugal
Tel. (351)(21) 7822400
Fax (351)(21) 7822443
E-mail internacional@
    cruzvermelha.org.pt

## Qatar Red Crescent Society

P.O. Box 5449
Doha
Qatar
Tel. (974)(4) 435111
Fax (974)(4) 439950
E-mail qrcs@qatar.net.qa

## Romanian Red Cross

Strada Biserica Amzei, 29
Sector 1
Bucarest
Romania
Tel. (40)(1) 6593385
Fax (40)(1) 3128452
Tlx 10531 romcr r

## The Russian Red Cross Society

Tcheryomushkinski Proezd 5
117036 Moscow
Russian Federation
Tel. (7)(095) 1265731
Fax (7)(095) 3107048
Tlx 411400 IKPOL SU

## Rwandan Red Cross

B.P. 425
Kigali
Rwanda
Tel. (250) 585446
Fax (250) 585449
Tlx 22663 CRR RW

## Saint Kitts and Nevis Red Cross Society

Red Cross House
Horsford Road
P.O. Box 62
Basseterre
Saint Kitts and Nevis
Tel. (1)(869) 4652584
Fax (1)(869) 4668129
E-mail skbredcr@caribsurf.com

## Saint Lucia Red Cross

Vigie
P.O. Box 271
Castries St Lucia, W.I.
Saint Lucia
Tel. (1)(758) 4525582
Fax (1)(758) 4537811
E-mail sluredcross@candw.lc

## Saint Vincent and the Grenadines Red Cross

HaliFax Street
Ministry of Education compound
Kingstown
P.O. Box 431
Saint Vincent and the Grenadines
Tel. (1)(784) 4561888
Fax (1)(784) 4856210
E-mail svgredcross@caribsurf.com

## Samoa Red Cross Society

P.O. Box 1616
Apia
Samoa
Tel. (685) 23686
Fax (685) 22676
E-mail samoaredcross@samoa.ws

## Red Cross of the Republic of San Marino

Via Scialoja, Cailungo
Republic of San Marino 47031
Tel. (37)(8) 994360
Fax (37)(8) 994360
E-mail crs@omniway.sm
Web http://www.tradecenter.sm/crs/

## Sao Tome and Principe Red Cross

Avenida 12 de Julho No. 11
B.P. 96
Sao Tome
Sao Tome and Principe
Tel. (239)(12) 22469
Fax (239)(12) 22305
E-mail cvstp@sol.stome.telepac.net

## Saudi Arabian Red Crescent Society

General Headquarters
Riyadh 11129
Saudi Arabia
Tel. (966)(1) 4740027
Fax (966)(1) 4740430
E-mail redcrescent@zajil.net

## Senegalese Red Cross Society

Boulevard F. Roosevelt
B.P. 299
Dakar
Senegal
Tel. (221) 8233992
Fax (221) 8225369

## Seychelles Red Cross Society

Place de la République
B.P. 53
Victoria
Mahé
Seychelles
Tel. (248) 324646
Fax (248) 321663
E-mail redcross@seychelles.net
Web http://www.
    seychelles.net/redcross/

## Sierra Leone Red Cross Society

6 Liverpool Street
P.O. Box 427
Freetown
Sierra Leone
Tel. (232)(22) 229384
Fax (232)(22) 229083
E-mail slrcs@sierratel.sl

## Singapore Red Cross Society

Red Cross House
15 Penang Lane
Singapore 238486
Tel. (65) 3360269
Fax (65) 3374360
E-mail redcross@cyberway.com.sg
Web http://www.redcross.org.sg/

## Slovak Red Cross

Grösslingova 24
814 46 Bratislava
Slovakia
Tel. (421)(7) 52923576
Fax (421)(7) 52923279
E-mail headq@redcross.sk

## Slovenian Red Cross

Mirje 19
P.O. Box 236
SI-61000 Ljubljana
Slovenia
Tel. (386)(1) 2414300
Fax (386)(1) 2414344
E-mail rdeci.kriz-slo@guest.arnes.si

## The Solomon Islands Red Cross

P.O. Box 187
Honiara
Solomon Islands
Tel. (677) 22682
Fax (677) 25299
E-mail sirc@solomon.com.sb

## Somali Red Crescent Society

c/o ICRC Box 73226
Nairobi
Kenya
Tel. (871 or 873) 131 2646
    (Mogadishu) / (254)(2) 723963
    (Nairobi)
Fax 1312647 (Mogadishu) /
    715598 (Nairobi)

## The South African Red Cross Society

1st Floor, Helen Bowden Bldg
Beach Road, Granger Bay
P.O. Box 50696, Waterfront
Cape Town 8002
South Africa
Tel. (27)(21) 4186640
Fax (27)(21) 4186644
E-mail sarcs@redcross.org.za

## Spanish Red Cross

Rafael Villa, s/n (Vuelta Ginés Navarro)
28023 El Plantio
Madrid
Spain
Tel. (34)(91) 3354444
Fax (34)(91) 3354455
E-mail informa@cruzroja.es
Web http://www.cruzroja.es/

## The Sri Lanka Red Cross Society

307, 2/1 T.B. Jayah Mawatha
P.O. Box 375
Colombo 10
Sri Lanka
Tel. (94)(1) 678420
Fax (94)(1) 695434
E-mail slrc@sri.lanka.net

## The Sudanese Red Crescent

Al Mak Numir St/Gamhouria St.
Plot No. 1, Block No 4
P.O. Box 235
Khartoum
Sudan
Tel. (249)(11) 772011
Fax (249)(11) 772877
E-mail srcs@sudanmail.net

## Suriname Red Cross

Gravenberchstraat 2
Postbus 2919
Paramaribo
Suriname
Tel. (597) 498410
Fax (597) 464780
E-mail surcross@sr.net

## Baphalali Swaziland Red Cross Society

104 Johnstone Street
P.O. Box 377
Mbabane
Swaziland
Tel. (268) 4042532
Fax (268) 4046108
E-mail bsrcs@redcross.sz

## Swedish Red Cross
Hornsgatan 54
Box 17563
SE-118 91 Stockholm
Sweden
Tel. (46)(8) 4524600
Fax (46)(8) 4524761
E-mail postmaster@redcross.se
Web http://www.redcross.se/

## Swiss Red Cross
Rainmattstrasse 10
Postfach
3001 Bern
Switzerland
Tel. (41)(31) 3877111
Fax (41)(31) 3877122
E-mail info@redcross.ch
Web http://www.redcross.ch/

## Syrian Arab Red Crescent
Al Malek Aladel Street
Damascus
Syrian Arab Republic
Tel. (963)(11) 4429662
Fax (963)(11) 4425677
E-mail SARC@net.sy

## Red Crescent Society of Tajikistan
120, Umari Khayom St.
734017, Dushanbe
Tajikistan
Tel. (992)(372) 240374
Fax (992)(372) 245378
E-mail rcstj@yahoo.com

## Tanzania Red Cross National Society
Ali Hassan Mwinyi Road, Plot 294/295
Upanga
P.O. Box 1133
Dar es Salaam
Tanzania, United Republic of
Tel. (7)(22) 2150881
Fax (7)(22) 2150147
E-mail redcrosstz.foc@raha.com

## The Thai Red Cross Society
Terd Prakiat Building
1871 Henry Dunant Road
Bangkok 10330
Thailand
Tel. (66)(2) 2564037
Fax (66)(2) 2553064
E-mail webmaster@redcross.or.th
Web http://www.redcross.or.th/

## Togolese Red Cross
51, rue Boko Soga
Amoutivé
B.P. 655
Lome
Togo
Tel. (228) 212110
Fax (228) 215228
E-mail crtogol@syfed.tg.refer.org

## Tonga Red Cross Society
P.O. Box 456
Nuku'Alofa
South West Pacific
Tonga
Tel. (676) 21360
Fax (676) 24158
E-mail redcross@kalianet.to

## The Trinidad and Tobago Red Cross Society
7A, Fitz Blackman Drive
Wrightson Road
P.O. Box 357
Port of Spain
Trinidad and Tobago
Tel. (1)(868) 76278215
Fax (1)(868) 76278215
E-mail ttrcs@carib-link.net

## Tunisian Red Crescent
19, Rue d'Angleterre
Tunis 1000
Tunisia
Tel. (216)(71) 862485
Fax (216)(71) 862971
E-mail hilal.ahmar@planet.tn

## Turkish Red Crescent Society
Atac Sokak 1 No. 32
Yenisehir
Ankara
Turkey
Tel. (90)(312) 4302300
Fax (90)(312) 4300175
Tlx 44593 KZLY TR
Web http://www.kizilay.org.tr/

## Red Crescent Society of Turkmenistan
48 A. Novoi str.
744000 Ashgabat
Turkmenistan
Tel. (993)(12) 395511
Fax (993)(12) 351750
E-mail nrcst @online.tm

## The Uganda Red Cross Society
Plot 28/30 Lumumba Avenue
P.O. Box 494
Kampala
Uganda
Tel. (256)(41) 258701
Fax (256)(41) 258184
E-mail sg.urcs@imul.com

## Ukrainian Red Cross Society
30, Pushkinskaya St.
252004 Kiev
Ukraine
Tel. (380)(44) 2350157
Fax (380)(44) 2351096
E-mail redcross@ukrpack.net
Web http://www.redcross.org.ua

## Red Crescent Society of the United Arab Emirates
P.O. Box 3324
Abu Dhabi
United Arab Emirates
Tel. (9)(712) 6419000
Fax (9)(712) 6420101
E-mail hilalrc@emirates.net.ae

## British Red Cross

9 Grosvenor Crescent
London SW1X 7EJ
United Kingdom
Tel. (44)(207) 2355454
Fax (44)(207) 2456315
E-mail information@
    redcross.org.uk
Web http://www.redcross.org.uk

## American Red Cross

431 18th Street NW, 2nd Floor
Washington, DC 20006
United States
Tel. (1)(202) 6393400
Fax (1)(202) 6393595
E-mail postmaster@usa.redcross.org
Web http://www.redcross.org/

## Uruguayan Red Cross

Avenida 8 de Octubre, 2990
11600 Montevideo
Uruguay
Tel. (598)(2) 4802112
Fax (598)(2) 4800714
E-mail cruzroja@adinet.com.uy

## Red Crescent Society of Uzbekistan

30, Yusuf Hos Hojib St.
700031 Tashkent
Uzbekistan
Tel. (988)(712) 563741
Fax (988)(712) 561801
E-mail rcuz@uzpak.uz
Web http://www.redcrescent.uz/

## Vanuatu Red Cross Society

P.O. Box 618
Port Vila
Vanuatu
Tel. (678) 27418
Fax (678) 22599
E-mail redcross@vanuatu.com.vu

## Venezuelan Red Cross

Avenida Andrés Bello, 4
Apartado 3185
Caracas 1010
Venezuela
Tel. (58)(2) 5714380
Fax (58)(2) 5761042
E-mail dirnacsoc@cantv.net

## Red Cross of Viet Nam

82, Nguyen Du Street
Hanoï
Viet Nam
Tel. (844)(8) 225157
Fax (844)(9) 424285
E-mail vnrchq@netnam.org.vn
Web http://www.vnrc.org.vn

## Yemen Red Crescent Society

Head Office, Building No. 10
26 September Street
P.O. Box 1257
Sanaa
Yemen
Tel. (967)(1) 283132 / 283133
Fax (967)(1) 283131
Tlx 3124 HILAL YE

## Yugoslav Red Cross

Simina 19
11000 Belgrade
Yugoslavia
Tel. (381)(11) 623564
Fax (381)(11) 622965
E-mail jckbg@jck.org.yu
Web http://www.jck.org/yu

## Zambia Red Cross Society

2837 Los Angeles Boulevard
Longacres
P.O. Box 50001 (Ridgeway 15101)
Lusaka
Zambia
Tel. (260)(1) 250607
Fax (260)(1) 252219
E-mail zrcs@zamnet.zm

## Zimbabwe Red Cross Society

98 Cameron Street
P.O. Box 1406
Harare
Zimbabwe
Tel. (263)(4) 775416
Fax (263)(4) 751739
E-mail zrcs@harare.iafrica.com

Section Two

**Tracking
the system**

International
Federation
delegations help
National Societies
alleviate the suffering
of vulnerable people.

International
Federation,
Afghanistan 2001.

# A global network

Contact details for regional and country delegations of the International Federation of Red Cross and Red Crescent Societies. Information correct as of 1 March 2002.

## International Federation of Red Cross and Red Crescent Societies
P.O. Box 372
1211 Geneva 19
Switzerland
Tel. 4122 7304222
Fax 4122 7330395
Tlx 045 412 133 FRC CH
Tlg. LICROSS GENEVA
E-mail secretariat@ifrc.org
Web http://www.ifrc.org

## Red Cross/European Union Office
Rue J. Stallaert 1, bte 14
1050 - Brussels
Belgium
Tel. 322 3475750
Fax 322 3474365
E-mail rceulb.brux@inforboard.be

## International Federation of Red Cross and Red Crescent Societies at the United Nations
630 Third Avenue
21st floor
Suite 2104
New York NY10017
United States
Tel. 1212 3380161
Fax 1212 3389832

## International Federation regional delegations

### Buenos Aires
Lucio V. Mansilla 2698 2°
1425 Buenos Aires
Argentina
Tel. 54114 9638659
Fax 54114 9613320
E-mail ifrcar01@ifrc.org

### Yaoundé
Rue Mini-Prix Bastos
BP 11507
Yaoundé
Cameroon
Tel. 237 2217437
Fax 237 2217439
E-mail ifrccm04@ifrc.org

### Beijing
Apt. 4-2-51 Building 4,
Entrance 2,
Floor 5, Apt.1
Jianguomenwai Diplomatic
Compound,
Chaoyang District,
Beijing 100600
China
Tel. 8610 65327162
Fax 8610 65327166

### Abidjan
II Plateaux Polyclinique-
Lot no. 14-Îlot
Villa Duplex
04 PO Box 2090,
Abidjan, 04
Côte d'Ivoire
Tel. 225 22404450
Fax 225 22404459
E-mail tfrc.gn01@ifrc.org

### Santo Domingo
Mustafa Kemal Atur, 21
Ensanche Naco
Santo Domingo
Dominican Republic
Tel. 1809 5673344
Fax 1809 5675395
E-mail ifrc01@codetel.net.do

### Suva
P.O. Box 2507
Government Building
Suva
Fiji
Tel. 679 311855
Fax 679 311406
E-mail ifrcrds@is.com.fj

### Guatemala City
Av. de las Américas
Pl. Uruguay
Ciudad de Guatemala 01014
Guatemala
Tel. 502 3335425
Fax 502 3631449
E-mail fedecruz@guate.net

### Budapest
Zolyomi Lepcso Ut 22
1124 Budapest
Hungary
Tel. 361 3193423
Fax 361 3193424
E-mail ifrchu01@ifrc.org

### New Delhi
F-25A Hauz Khas Enclave
New Delhi 110 016
India
Tel. 9111 6858671
Fax 9111 6857567
E-mail ifrcin01@ifrc.org

### Amman
Al Shmeisani
Maroof Al Rasafi Street
Building No. 19
P.O. Box 830511 / Zahran
Amman,
Jordan
Tel. 962 6 5681060
Fax 962 6 5694556
E-mail ifrcjo01@ifrc.org

### Almaty
86, Kunaeva Street
480100 Almaty
Kazakhstan
Tel. 732 72 918838
Fax 732 72 914267
E-mail ifrckz01@ifrc.org

### Nairobi
Woodlands Road
P.O. Box 412275
Nairobi,
Kenya
Tel. 254 2 712159
Fax 254 2 718415
E-mail ifrcke01@ifrc.org

### Bangkok
18th Floor, Ocean Tower 2
75/26 Sukhumvit 19
Wattana
Bangkok 10110
Thailand
Tel. 662 6616933
Fax 662 6616937
E-mail ifrcth02@ifrc.org

### Harare
9, Coxwell Road
Milton Park
Harare
Zimbabwe
Tel. 2634 720315
Fax 2634 708784
E-mail ifrczw01@ifrc.org

## International Federation country delegations

### Afghanistan
43D S. Jamal-ud-Din Afghani Rd.
University Town
Peshawar
Pakistan
Tel. 873 382280530
Fax 873 382280534
E-mail kabul@wireless.ifrc.org

### Albania
c/o Albanian Red Cross
Rruga "Muhamet Gjollesha"
Sheshi "Karl Topia"
P.O. Box 1511
Tirana,
Albania
Tel. 355 4256708
Fax 355 4227966
E-mail irfca102@ifrc.org

### Angola
Caixa Postal 3324
Rua 1o Congresso nr. 27
Luanda
Angola
Tel. 2442 372868
Fax 2442 372868
E-mail ifrcao01@ifrc.org

### Armenia
Gevorg Chaush St. 50/1
Yerevan 375088
Armenia
Tel. 3741 354649
Fax 3741 395731
E-mail ifrcam03@ifrc.org

### Azerbaijan
Qizil Xac/Qizil Aypara Evi
First Magomayev Street 6A
370004 Baku
Azerbaijan
Tel. 99412 925792
Fax 99412 971889
E-mail baku02@ifrc.org

### Bangladesh
c/o Bangladesh Red Crescent
Society
684-686 Bara Magh Bazar
Dhaka - 1217
Bangladesh
Tel. 8802 8315401
Fax 8802 9341631
E-mail ifrcbd@citecho.net

### Belarus
Ulitsa Mayakovkosgo 14
Minsk 220006
Belarus
Tel. 375172 217273
Fax 375172 219060
E-mail ifrcby01@ifrc.org

### Bosnia and Herzegovina
Titova 7
71000 Sarajevo
Bosnia and Herzegovina
Tel. 38733 666009
Fax 38733 666010
E-mail ifrcbih01@ifrc.org

## Burundi
Avenue des États-Unis 3674A
B.P. 324
Bujumbura
Burundi
Tel. 257 229524
Fax 257 229408

## Cambodia
17 Deo, Street Croix-Rouge
Central Post Office/P.O. Box 620
Phnom Penh
Cambodia
Tel. 85523 210162
Fax 85523 210163
E-mail ifrckh01@ifrc.org

## Colombia
c/o Colombian Red Cross
Avenida 68 N° 66-31
Bogotá
Colombia
Tel. 571 4299329
Fax 571 4299328
E-mail
ifrcbogota@sncruzroja.org.co

## Congo, Democratic Republic of the
21 Avenue Flamboyant
Place de Sefoutiers
Gombe
Kinshasa
Democratic Republic of the Congo
Tel. 243 1221495
Fax. 871 682 049 073
E-mail: hod@ficr.aton.cd

## Congo, Republic of
60, Av. de la Libération de Paris
B.P. 88
Brazzaville
Republic of Congo
Tel. 242 667564
Fax. 242 815044

## East Timor
c/o ICRC delegation
Bairro dos Grilhos 20
P.O. Box 73
Dili
East Timor
Tel. 670 321448
Fax 670 390322899

## El Salvador
c/o Salvadorean Red Cross Society
17 Calle Pte. y Av. Henri Dunant
Apartado Postal 2672
San Salvador
El Salvador
Tel. 503 8902134
Fax 503 2811932
E-mail ifrcsv11@ifrc.org

## Eritrea
c/o Red Cross Society of Eritrea
Andnet Street
P.O. Box 575
Asmara
Eritrea
Tel. 2911 150550
Fax 2911 151859
E-mail ifrc@eol.com.er

## Ethiopia
Ras Destra Damtew Avenue
P.O. Box 195
Addis Ababa
Ethiopia
Tel. 2511 514571
Fax 2511 512888
E-mail ifrcet04@ifrc.org

## Georgia
7, Anton Katalikosi Street
Tbilisi
Georgia
Tel. 99532 950945
Fax 99532 985976
E-mail ifrcge01@ifrc.org

## Guatemala
c/o Cruz Roja Guatemalteca
3a Calle 8-40, Zona 1
Ciudad de Guatemala
Guatemala
Tel. 502 2532809
Fax 502 2200672
E-mail fedecng@intelnet.net.gt

## Guinea
Coleah, route du Niger
Près de l'Ambassade
de Yougoslavie
B.P. 376
Conakry
Guinea
Tel. 224 413825
Fax 224 414255
E-mail ifrc.gn01@ifrc.org

## Honduras
Colonia Florencia Norte segunda
calle casa No. 1030
contigua al edificio Tovar Lopez
Tegucigalpa
Honduras
Tel. 504 2320707
Fax 504 2390718

## Indonesia
c/o Indonesian Red Cross Society
P.O. Box 2009
Jakarta
Indonesia
Tel. 6221 79191841
Fax 6221 79180905
E-mail ifrcid01@ifrc.org

## Iraq
c/o Iraqi Red Crescent Society
P.O. Box 6143
Baghdad
Iraq
Tel. 964 1 5370042
Fax 964 1 5372547

## Korea, DPR

c/o Red Cross Society of the DPR
Korea
Ryonwa 1, Central District
Pyongyang
Korea, Democratic People's
    Republic of
Tel. 8502 3813490
Fax 8502 3813490

## Laos

c/o Lao Red Cross
P.O.Box 2948
Setthatirath Road, Xiengnhune
Vientiane
Lao People's Democratic Republic
Tel. 856 21215762
Fax 856 21215935
E-mail laoifrc@laotel.com

## Lebanon

N. Dagher Building
Mar Tacla
Beirut
Lebanon
Tel. 9611 424851
Fax 9615 459658
E-mail ifrclb01@ifrc.org

## Macedonia, FYR of

Bul. Koco Racin 13
Skopje 9100
Macedonia, Former Yugoslav
Republic of
Tel. 3892 114271
Fax 3892 115240
E-mail ifrcmk01@ifrc.org

## Mongolia

c/o Red Cross Society of Mongolia
Central Post Office
Post Box 537
Ulaanbaatar
Mongolia
Tel. 97611 320171
Fax 97611 321684
E-mail ifrcmongol@magicnet.mn

## Mozambique

Av. 24 de Julho 641
Caixa postal 2488
Maputo
Mozambique
Tel. 2581 303959
Fax 2581 492278
E-mail ifrcmz01@ifrc.org

## Myanmar

c/o Myanmar Red Cross Society
42 Strand Road
Yangon
Myanmar
Tel. 951 297877
Fax 951 297877
E-mail ifrc@mptmail.net.mm

## Nicaragua

c/o Nicaraguan Red Cross
Reparto Belmonte, Carretera Sur
Apartado Postal P-48 Las
Piedrecitas
Managua
Nicaragua
Tel. 505 2650186
Fax 505 2652069

## Nigeria

c/o Nigerian Red Cross Society
11, Eko Akete Close
Off St. Gregory's Road
South West Ikoyi
P.O. Box 764
Lagos
Nigeria
Tel. 2341 2695228
Fax 2341 2695229
E-mail fedcross@infoweb.abs.net

## Pakistan

c/o Pakistan Red Crescent Society
National Headquarters
Sector H-8
Islamabad
Pakistan
Tel. 9251 9257122
Fax 9251 4430745
E-mail ifrcaf@isb.comsats.net.pk

## Papua New Guinea

c/o PNG Red Cross
P.O. Box 6545
Boroko
Papua New Guinea
Tel. 675 3112277
Fax 675 3230731

## Peru

c/o Peruvian Red Cross
Av. Arequipa 1285
Lima
Peru
Tel. 511 4700599
Fax 511 470 0606
E-mail cruzrojaaqp3@unsa.edu.pe

## Russian Federation

c/o Russian Red Cross Society
Tcheryomushkinski Proezd 5
117036 Moscow
Russian Federation
Tel. 7502 9375267
Fax 7502 9375263
E-mail moscow@ifrc.org

## Rwanda

c/o Rwandan Red Cross
B.P. 425, Nyamiranbo
Kigali
Rwanda
Tel. 250 8470452
Fax 250 73233
E-mail ifrcrw01@ifrc.org

## Sierra Leone

c/o Sierra Leone Red Cross Society
6, Liverpool Street
P.O. Box 427
Freetown
Sierra Leone
Tel. 23222 227772
Fax 23222 228180
E-mail ifrc@sierratel.sl

## Somalia

Chaka Road
off Argwings Kodhek Road
P.O. Box 41275
Nairobi
Kenya
Tel. 2542 728294
Fax 2542 729070
E-mail ifrcs001@ifrc.org

## Sri Lanka

3rd floor, 307 T B Jayah Mawatha
LK Colombo
Sri Lanka
Tel. 9474 7155977
Fax 9474 571275
E-mail ifrclk01@srilanka.net

## Sudan

Al Mak Nimir Street/Gamhouria
Street
Plot No 1, Block No 4
P.O. Box 10697
East Khartoum
Sudan
Tel. 24911 771033
Fax 24911 770484

## Tajikistan

c/o Tajikistan Red Crescent Society
120, Omar Khayom St.
734017 Dushanbe
Tajikistan
Tel. 992372 245981
Fax 992372 248520
E-mail ifrcdsb@ifrc.org

## Tanzania

Ali Hassan Mwinyi Road
Plot No. 294/295
P.O. Box 1133
Dar es Salaam
Tanzania, United Republic of
Tel. 255 22 21116514
Fax 255 21117308
E-mail ifrctz01@ifrc.org

## Turkey

Ziya Gökalp Caddesi
Adakale Sokak no. 27/8
TR-06420 Ankara
Turkey
Tel. 90312 4357728
Fax 90312 4359896

## Uganda

c/o Uganda Red Cross Society
Plot 97, Buganda Road
P.O. Box 494
Kampala
Uganda
Tel. 25641 234968
Fax 25641 258184
E-mail ifrc@imul.com

## Viet Nam

19 Mai Hac De Street
Hanoï
Viet Nam
Tel. 844 9438250
Fax 844 9436177
E-mail ifrc@hn.vnn.vn

## Yugoslavia, Federal Republic of

Simina Ulica Broj 21
11000 Belgrade
Yugoslavia, Federal Republic of
Tel. 381 11 3282202
Fax 381 11 3281791
E-mail telecom@ifrc.org.yu

## Zambia

c/o Zambia Red Cross
2837 Los Angeles Boulevard
P.O. Box 50001
Ridgeway 15101
Lusaka
Zambia
Tel. 2601 254074
Fax 2601 261878
E-mail ifrczmb03@ifrc.org

# Index